MW00366656
G-AGPE

Winkle

Winkle

The Extraordinary Life of Britain's Greatest Pilot

PAUL BEAVER

PENGUIN MICHAEL JOSEPH

UK | USA | Canada | Ireland | Australia
India | New Zealand | South Africa

Penguin Michael Joseph is part of the Penguin Random House group of companies
whose addresses can be found at global.penguinrandomhouse.com

First published 2023
001

For picture permissions see page 495

Set in 13.75/18pt Garamond MT Std
Typeset by Jouve (UK), Milton Keynes
Printed and bound in Great Britain by Clays Ltd, Elcograf S.p.A.

The authorized representative in the EEA is Penguin Random House Ireland,
Morrison Chambers, 32 Nassau Street, Dublin D02 YH68

A CIP catalogue record for this book is available from the British Library

ISBN: 978–0–718–18670–8

www.greenpenguin.co.uk

Penguin Random House is committed to a sustainable future for our business, our readers and our planet. This book is made from Forest Stewardship Council® certified paper.

To Eric for a deep friendship of over forty years and
for allowing me to write this truly amazing story

Contents

PART 3

Test Pilot 1942–49

PART 4

Full Career 1949–70

PART 5

Introspection and Reflection

Maps

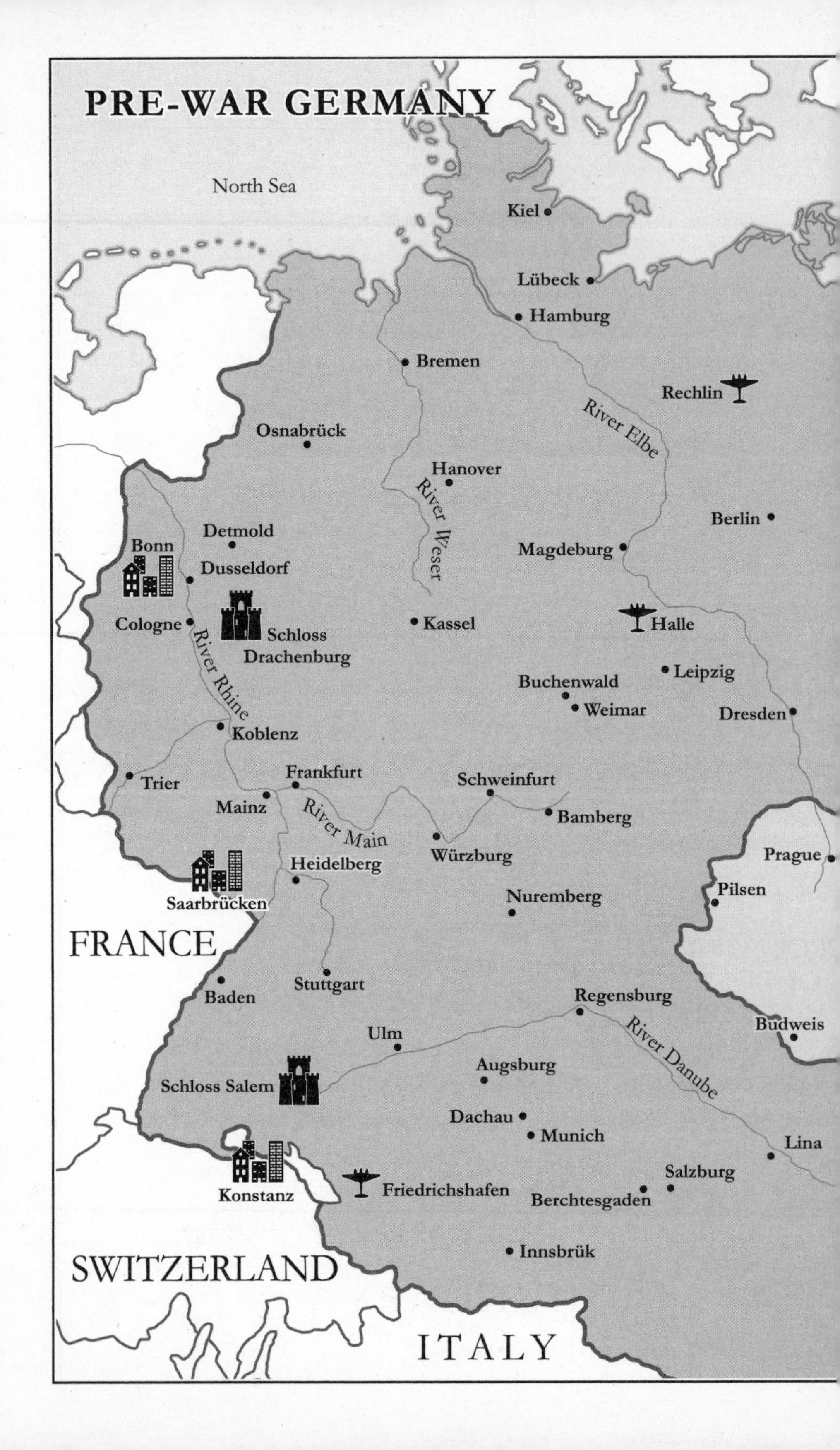
PRE-WAR GERMANY
North Sea
Kiel
Lübeck
Hamburg
Bremen
Rechlin
River Elbe
Osnabrück
Hanover
River Weser
Berlin
Detmold
Bonn
Dusseldorf
Magdeburg
Cologne
Schloss
Drachenburg
Kassel
Halle
River Rhine
Leipzig
Buchenwald
Weimar
Dresden
Koblenz
Trier
Frankfurt
Schweinfurt
Mainz
Bamberg
River Main
Würzburg
Prague
Heidelberg
Pilsen
Saarbrücken
Nuremberg
FRANCE
Baden
Stuttgart
Regensburg
Budweis
Ulm
River Danube
Augsburg
Schloss Salem
Dachau
Munich
Lina
Salzburg
Konstanz
Friedrichshafen
Berchtesgaden
Innsbrük
SWITZERLAND
ITALY

Memel
N
Baltic Sea
Königsberg
Danzig
ettin
Thorn
POLAND
Posen
Vistula River
Warsaw
Lödz
River Oder
Breslau
Kraków
PROTECTORATE
OF BOHEMIA
AND MORAVIA
Brno
SLOVAKIA
Vienna
Bratislava
0
200 miles
0
300 km
HUNGARY
az
Budapest

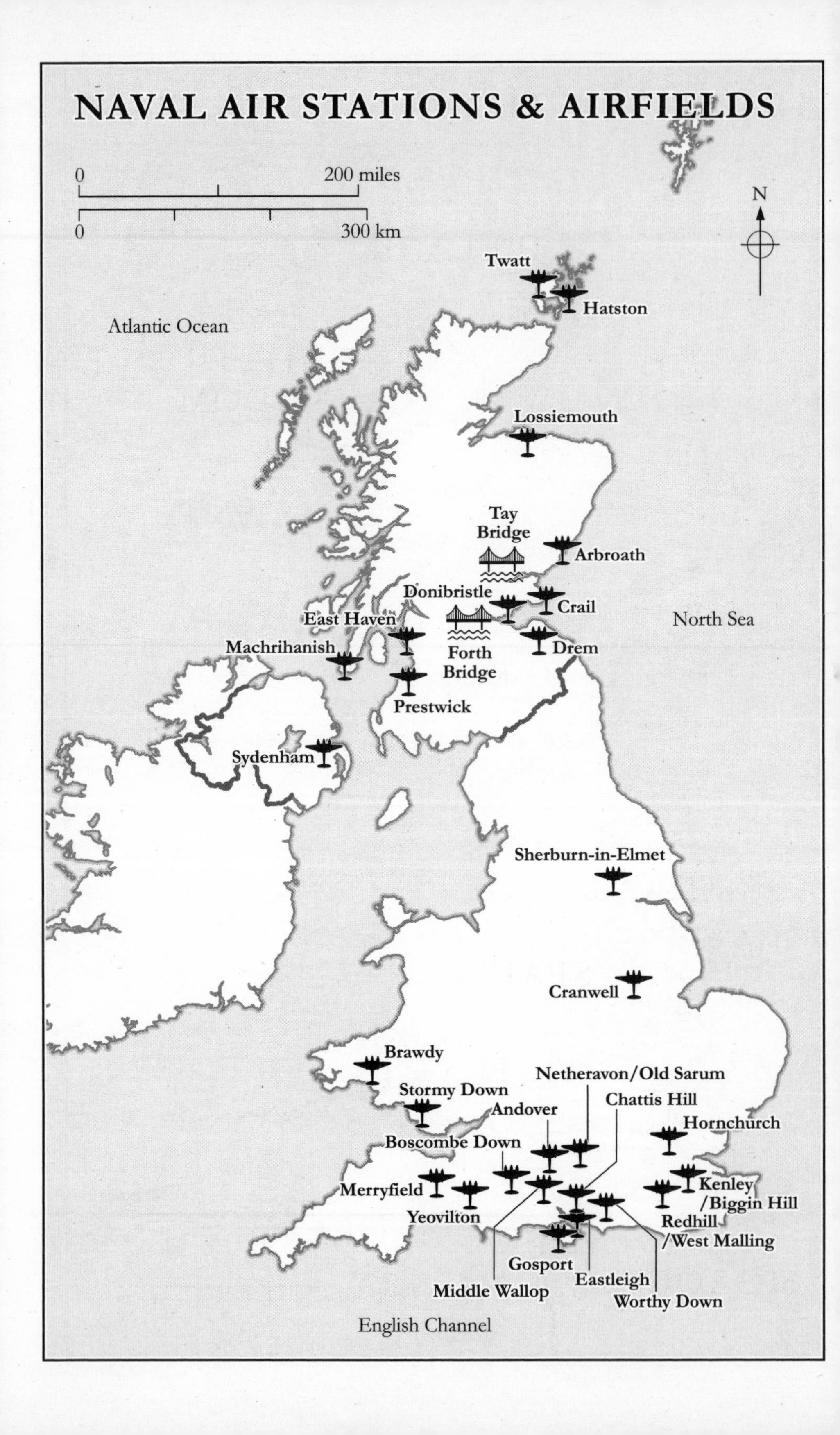
NAVAL AIR STATIONS & AIRFIELDS
0
200 miles
0
300 km
N
Atlantic Ocean
Twatt
Hatston
Lossiemouth
Tay Bridge
Arbroath
Donibristle
Crail
East Haven
Machrihanish
Forth Bridge
Drem
North Sea
Prestwick
Sydenham
Sherburn-in-Elmet
Cranwell
Brawdy
Netheravon/Old Sarum
Stormy Down
Chattis Hill
Andover
Hornchurch
Boscombe Down
Merryfield
Kenley /Biggin Hill
Yeovilton
Redhill /West Malling
Gosport
Eastleigh
Middle Wallop
Worthy Down
English Channel

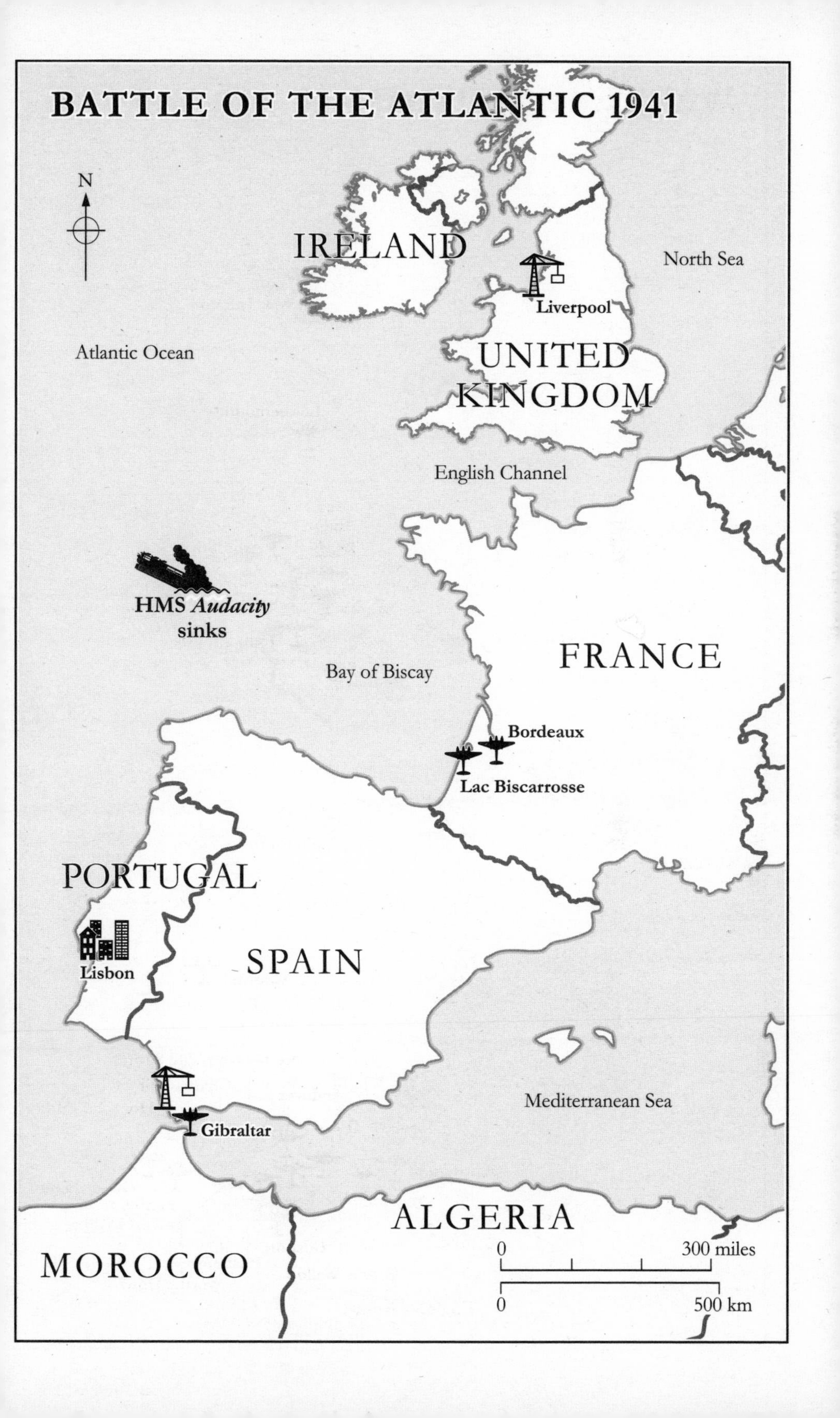
BATTLE OF THE ATLANTIC 1941
N
IRELAND
North Sea
Liverpool
Atlantic Ocean
UNITED
KINGDOM
English Channel
HMS *Audacity*
sinks
FRANCE
Bay of Biscay
Bordeaux
Lac Biscarrosse
PORTUGAL
SPAIN
Lisbon
Mediterranean Sea
Gibraltar
ALGERIA
0
300 miles
MOROCCO
0
500 km

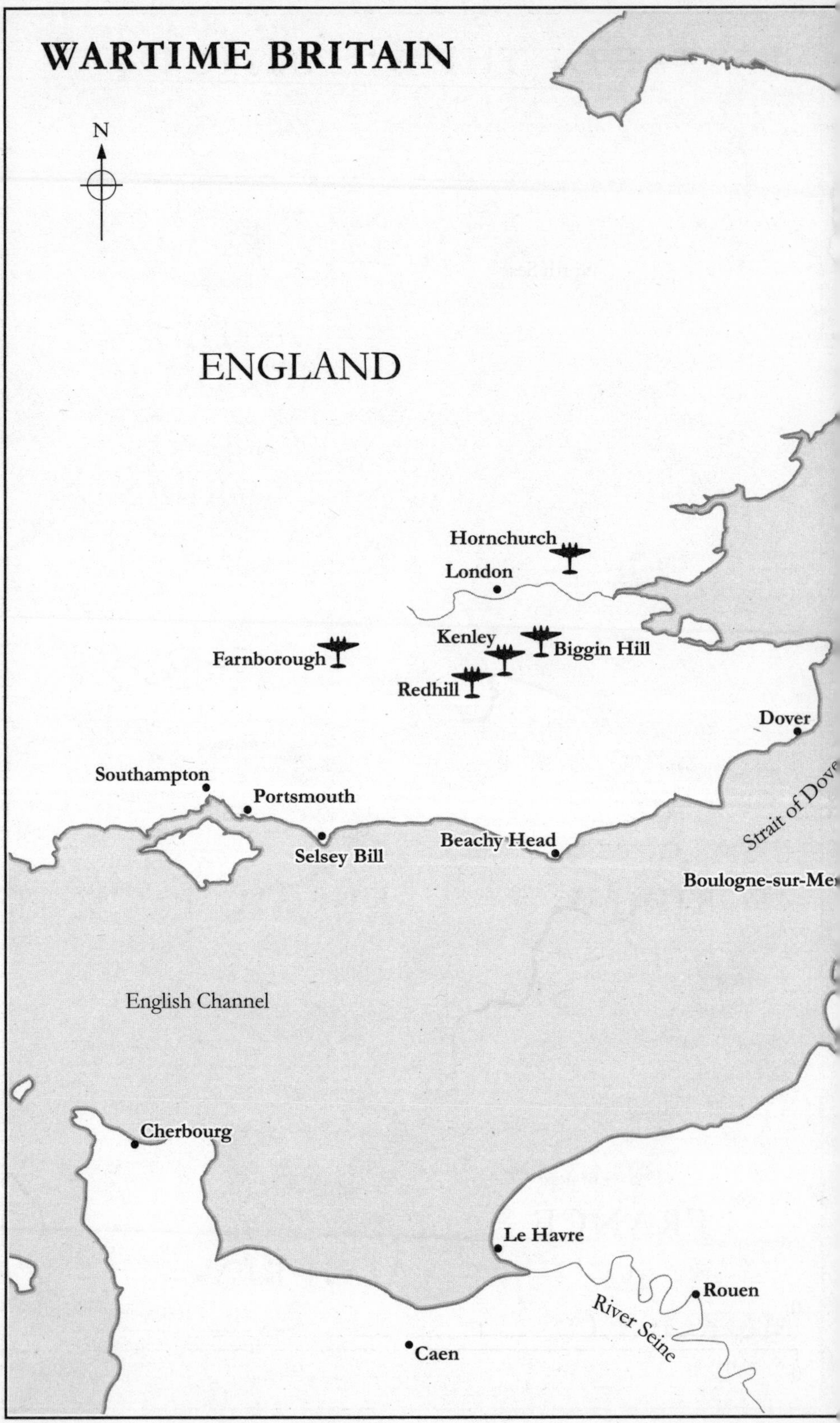

WARTIME BRITAIN

N

ENGLAND

Hornchurch
London
Kenley
Biggin Hill
Farnborough
Redhill
Dover
Southampton
Portsmouth
Strait of Dove
Selsey Bill
Beachy Head
Boulogne-sur-Me
English Channel
Cherbourg
Le Havre
Rouen
River Seine
Caen

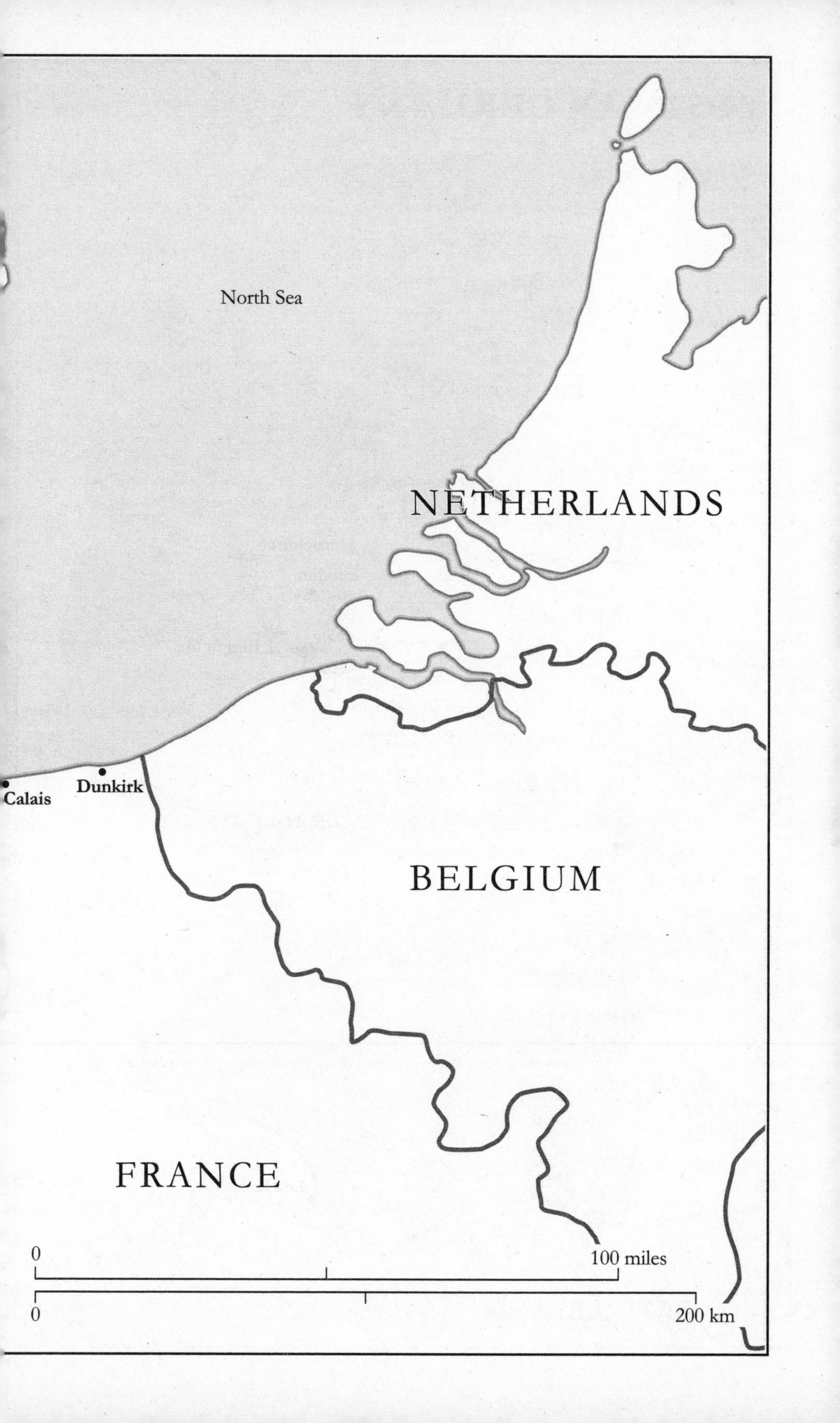
North Sea
NETHERLANDS
Calais
Dunkirk
BELGIUM
FRANCE
0
100 miles
0
200 km

POSTWAR GERMANY
N
DENMARK
Baltic Sea
North Sea
Leck
Flensburg
Schleswig
Kiel
Jagel
Plön
Tarnewitz
Peenemünde
Nordholz
Lübeck
Rechlin
Barnstedt
Fassberg
Bergen-Belsen
Berlin
Gatow
Celle
BRITISH SECTOR
SOVIET SECTOR
Grossenhain
Kassel
CZECHOSLOVAKIA
LUXEMBOURG
AMERICAN SECTOR
FRENCH SECTOR
SAAR PROTECTORATE (FRANCE)
Largerlechfeld
Benningen
Salzburg
AUSTRIA
0
100 miles
0
200 km
SWITZERLAND

Winkle

Eric Brown in his office, his beloved Martlet fighter. In this robust American naval fighter he carried out the first of 2,407 deck landings, downed at least three enemy aircraft, attacked U-boats on the surface and survived a devastating counter-attack from a Condor bomber, resulting in a harrowing deck landing. The prominent gunsight was a frequent hazard when Eric's Martlet crashed, and he took to the grave a part of one, embedded in his jaw.

Prologue

October 1941: The Bay of Biscay

In the middle of the Atlantic Ocean, the British aircraft carrier HMS *Audacity* was a whirl of motion. In the pouring rain, two pilots raced across the lurching deck and jumped into the cockpits of their grey and green Martlet fighter planes. They knew the drill by heart. They also knew that the Germans would stop at nothing to sink the merchant ships in the convoy sailing from Gibraltar and needed to be stopped at all costs. This time it was bombers and, most likely, there were submarines in the area too.

In his cockpit, 21-year-old Eric Melrose Brown meticulously ran through his last-minute checks. A tug on his seat harness across his lap, to make sure it was tight, then a scan of the flight instruments, in a routine clockwise motion, switching on the tiny pin-prick navigation lamps, priming the engine and adjusting the fuel pump. He locked the canopy open. Unlike many of his fellow pilots, Eric left his canopy open for take-off, just in case the engine failed, and he crashed into the churning waves below. He had seen another pilot sink below the waves, frantically banging against a jammed canopy, before heading to the abyss. If anything was going to get him killed, it would be fighting in action, not an accident because of a faulty engine. And his engine started faultlessly. Not for the first time, he mentally thanked the squadron's mechanics who had worked through the night, in freezing conditions, to ensure the Martlets were serviceable.

Lining his Martlet up on the flight deck, he looked ahead as his leader, Roderick 'Sheepy' Lamb, roared down the deck and soared above the swirling ocean and into the grey sky. Suddenly, the signal up ahead on the bridge flashed green. *Go! Go! Go!* It was now Eric's turn. Opening the throttle to war emergency power, the supercharged Twin Wasp radial engine thundered in response, proceeding to hurtle the fighter forwards.

With its grey nose piercing the air, and the canopy open, Eric felt the full force of the wind and rain whip past his face, and then he was airborne. With his hand firmly on the throttle, Eric's stomach lurched as the Martlet momentarily sank below the level of the flight deck before climbing away, leaving the carrier and the ocean behind.

Soaring into the clouds, with the fighter's great

Hamilton Standard propeller in fine pitch[1] to make the most of the moist air, Eric finally slid the canopy shut ready for action and then turned the manual crack[2] thirty times, tucking the undercarriage into the fuselage.

Still climbing at more than 1,000 feet per minute, and breaking through the first layer of cloud, Eric felt rather than saw the enemy. His body involuntarily tensed. This only meant one thing. The enemy was close. Lamb waggled his wings to confirm the sighting of a German Condor bomber: 'Tally ho!'

The thick cloud suddenly parted, and the fearsome German aircraft came into view. In an instant, Eric and Lamb separated to split the bomber's defensive fire, with Lamb going down and right, and Eric deciding to go around to the port quarter, which intelligence said was the least defended. But the intelligence was wrong. The Condor was a flying porcupine, with dangerous weapons facing in every direction.

Just as Eric prepared to open fire, there was a thundering crash as the plexiglass of his side windscreen exploded. It took a moment for Eric to realize that he had been hit, and his mouth was full of blood. He

1 Fine pitch – and coarse pitch – are descriptions of the angle between the propeller section and the axle, which can be adjusted for optimum performance on take-off or for cruising etc.
2 A handle on the side of the cockpit.

was choking and coughing, with chunks of plexiglass jammed into his cheek, and the sky around him began to swirl into darkness.

From down and right, having seen the gunfire, Lamb quickly circled round, coming alongside Eric's starboard wing. Immediately, he saw the damage to the cockpit glass, and Eric slumped across the side of the canopy. He had passed out, while blood was splattered all over the cockpit. Unable to make contact, due to radio silence rules, Lamb could only watch in horror, praying that Eric would come round soon. If he didn't, then the Martlet, with Eric Brown in it, would surely plunge into the freezing Atlantic Ocean below. Watching on, his hand resting on the throttle, ready to pull away, Lamb suddenly saw some movement in the cockpit.

With a searing pain in his head, and coughing up blood, Eric pulled his body up straight. He was conscious, just. Groggily turning his head, he saw Lamb alongside him, frantically gesturing. Spitting out blood and plexiglass, he knew he needed to get back to the carrier if he was to survive, but the world again began to swirl, turning black, consciousness drifting away. Moments later, the Martlet began to drop to the ocean below . . .

21 January 2009: London

Upon entering the Army and Navy Club in St James', I made my way to the Nelson Room, where a small dinner party was due to take place. Our host, Captain Mike Nixon, the chief executive of the Fly Navy Heritage Trust, had called together the great and good of Britain's aviation industry in order to celebrate Eric 'Winkle' Brown's ninetieth birthday, a man whom most acknowledged as Britain's greatest ever pilot. I felt honoured to be invited.

By this stage I had known Eric for a number of years. We had first met in 1978, when, as a young man, I was veering between a sensible career in mining engineering and estate management and following my heart and writing a book on the aircraft carrier HMS *Ark Royal*. Eric was extremely helpful, not only contributing to the book but also convincing me to follow my passion. I therefore credited Eric for my subsequent career as an aviation journalist and author, and we had remained friends throughout.

Over dinner, the dozen or so guests drank champagne, and the conversation between fellow pilots and aviation historians burst into life, most of it focused on the daring deeds of our guest of honour. At the head of the table, Eric happily regaled us with some of his tales in his cultured Scottish brogue, still going strong after all these years since his Edinburgh upbringing. Of particular interest, of course, was how on earth he had managed to survive, after being injured by the German Condor and then falling unconscious. Indeed, many of the other heroic stories he shared from his subsequent time as a renowned test pilot during the war, and beyond, had morphed into legend. And yet, while many of the tales were already well known to us Eric Brown aficionados, such as his encounters with Hitler, Himmler, the German fighter aces and Neil Armstrong, others were real bombshells.

At the end of the night, when Eric's car arrived to take him back to Surrey, we all said a fond farewell and decided to continue the celebrations in his absence. Over the remaining port there was still much excitement surrounding the stories Eric had told. Phill O'Dell, one of Britain's finest test pilots, was particularly enthusiastic. He felt that Eric's story should be in print, where the full extent could be researched, explained and examined. Everyone agreed. He was clearly a national icon, who had not as yet received his due. His autobiography, first released in 1961, had also missed out a lot, before and after. At this Phill turned to me: 'You know Winkle well. You should write it.' A seed was sown.

The following week, Eric called to express yet again his thanks for the birthday party in London. During our conversation, I casually asked, 'What do you think about the possibility of me writing your biography?' The reply was immediate: 'Of course, dear chap. I will make sure you have access to the papers and letters.' But there was a caveat. 'But you can only write it when I am gone.' I was confused. Surely Eric wanted to receive recognition for his deeds while he was alive, but he stood firm. 'I only want a book about me to be released when I am dead.' That was the end of the matter.

When Eric sadly passed away in 2016, as promised his archives were presented to me. Among them were over twenty chests of documents, flight records, photo albums and memorabilia. It was a real treasure trove. Far more than I was expecting. Over the following few years, I eagerly went through the lot.

As expected, many of the papers backed up the incredible stories he had told in print and during his ninetieth birthday party. Some revealed yet even more astounding stories that he had chosen to keep to himself. However, some of the papers didn't always match with some of Eric's tales and required further investigation. All of this was probably to be expected. In any story there are elements of embellishment or poor recollection, usually due to advancing age. And yet some stories couldn't be so easily explained. Buried among this haul of documents were some incidents that truly shocked me. When investigating them, I finally realized why Eric had felt they were best left to be revealed until after he was gone. For right from the start of his life, there was a story that Eric had kept a secret until his death, even from his wife and his only son, and yet this was the story that really formed Eric Brown into the man he was to become.

PART I

The Lost Boy 1920–39

Eric Brown was the adored foster son of Robert and Euphemia Melrose Brown of Leith, near Edinburgh. Unable to have children themselves, they lavished attention and opportunity on the young boy from Hackney, creating a stable and loving home for him in Scotland.

1

Foundling

It was an unseasonably cold morning when the overnight express from London pulled into Waverley Station, Edinburgh, on 8 May 1920. The train that day included a carriage chartered by the National Children Adoption Association (NCAA), full of very young children needing homes. While London had no foster homes for these unwanted children, Edinburgh was ready to welcome them with open arms.

The *Scotsman*, Scotland's newspaper of record, later declared that '150 offers had been received from Scotland of homes for "unwanted" babies . . . one in Leith.' The said couple from the northern suburb of Edinburgh were 43-year-old Robert John and 42-year-old Euphemia Melrose Brown.

The Browns were devout lower-middle-class Presbyterians who had been married since 1910. Euphemia suffered from a club foot and wore Edwardian style long dresses to hide her elevated boot; Robert was a journeyman tailor, whose business had been interrupted by the First World War. When he had enlisted in 1916, he was thirty-nine years old, nominally too old to be posted to France as a front-line soldier. Despite this, he was so desperate to do his bit that he told the recruiter he was a year younger, with a convenient smudge over his birth certificate covering his birth date.

After just weeks in the front-line trenches, working as a member of the Labour Corps, rather than a front-line soldier, Robert developed 'trench foot' and was eventually sent back to 'Blighty' for more concentrated treatment. At forty years old, he clearly felt it was a young man's war on the front line and therefore took the opportunity to transfer to the Royal Flying Corps as a labourer on observation balloons, then transferring to the Royal Air Force when it was formed on 1 April 1918.

With the war over, Robert returned to Euphemia, and the couple's attention turned to having a family of their own. When nature deprived them of their greatest wish, fate decreed that Euphemia saw an advertisement to adopt young children from London. So, on this May morning, Robert and Euphemia had travelled across town to Waverley Station, carrying with them the necessary references from the church and a business connection, probably the local bank manager, hoping they would find a child to complete their family.

Wrapped up to keep out the chill, and holding each other's hands, the Browns craned their necks in anticipation as the train, and the NCAA carriage with it, came puffing into view. As the children jumped off the carriage and were allocated to their various families, Euphemia heard a baby cry. Before she set eyes on it, she decided there and then that she must give this child a home.

Handed a swaddle of blankets by a nurse, Euphemia caught sight of the crying baby boy for the first time. She was told his name was Eric, and that he had been born to a single mother called Dorothy on 21 January in the Salvation Army's Mothers' Hospital in Hackney. No reason was given for Dorothy giving up her child, but from that day forth, Euphemia and Robert were determined to give this abandoned crying boy a loving and secure home.

Despite this incredible start to life, this is a story that Eric took to the grave. It was never discussed in his autobiography, nor in any of his talks with friends or speeches. Not even his late wife Lynn or his only son Glenn were aware of it. He spent his whole life as a proud Scot, even though he was actually born in England. He even went as far as to keep a forged birth certificate in his papers to hide the true story of his origins. However, as we shall see, this was far from the only major discrepancy in Eric's 'approved' version of his life.

From the Browns' two-storey apartment at 269 Leith Walk, Edinburgh, Robert spent his days working as a tailor in the workshop on the ground floor. Yet it was his adopted father's occupation as a Royal Air Force flight lieutenant which really captured Eric's imagination. 'Somehow,' he says in his autobiography, 'I felt flying was in my blood.'

At first glance, this certainly seems to be the case. In addition to Robert's stories about flying during the War, he listed his occupation on Eric's birth certificate as 'Flight Lieutenant, Royal Air Force', and then on Eric's marriage certificate[1] as a 'retired RAF officer', with the rank of 'Squadron Leader'. Moreover, Robert's death certificate, issued when he passed away in December 1947, also recorded that he was a 'Retired Squadron Leader'.

In his autobiography Eric elaborated on this, talking excitedly of his father's 'flying training' and the 'high risk game' of being a pilot during the war, and recalled that a picture of Robert in Royal Flying Corps uniform was hung on their lounge wall. He also repeatedly referenced his first flight, when Robert took him flying 'on his knee' in a small fighter from RAF Turnhouse. These were stories Eric turned to not only in his autobiography, but

1 Eric's age on this official document is also in error. It quotes him as being twenty in 1942, although there is a convenient smudge on the age column. There is a stiff warning at the base of the document which states: 'making any alteration in this document constitutes an offence under the Forgery Act 1913 and is punishable accordingly'.

also time and time again in talks when discussing those who had inspired him. And yet none of it can be true.

Firstly, there is not a shred of evidence to back up any of Robert's claims. An inspection of his service records shows that while Robert did serve in the Royal Flying Corps during the war it was very much in a supporting role. He was not a Royal Flying Corps observer, or even a pilot-in-training; instead he likely helped prepare new sites in France for No. 14 Balloon Section, probably stringing out the telephone lines that were used for passing information directly from the gas-filled barrage balloons, which could oversee the battlefield, to forward artillery positions.

It is curious why Robert chose to invent such a lie. Perhaps it was to make his uneventful life seem more exciting and prestigious. Maybe it was to help embellish his war record, so that he could say that he truly served and played his part. Yet not only did Robert invent his fictitious flying career and persona, at some point Eric must have known it was all false, especially by the time he wrote the second edition of his autobiography in 2006, where many of these details were mentioned for the first time.

When it came to Eric reminiscing about flying with his father for the first time in a British fighter, his experience and knowledge would have told him that not only did a Gloster Gauntlet not serve the Auxiliary Air Force in Edinburgh, but it had not even entered service at this time. Perhaps, Robert and Eric were taken on a joyride, during a station home day, in an Avro 504K biplane trainer of No. 603 (City of Edinburgh Squadron) and the hazy memory made the event feel far more momentous. But why did Eric choose to perpetuate his father's lie that he was a decorated pilot, as well as keep his adoption a total secret? The reasons may become a little clearer later in our story.

Despite these inconsistencies, we know Eric was enrolled in the local state-funded Lorne Street Primary School, just off Leith Walk. It seems Eric was happy at school, where he blossomed at around ten years of age and was rated 'excellent' across the board.[2] He even won a much-coveted scholarship to Edinburgh's Royal High School[3] (RHS), which boasted Alexander Graham Bell and Sir Walter Scott among its alumni.

He had hoped to join the school for the 1930–31 year, but that autumn he contracted rubella,[4] which was sweeping through the schools of Britain, especially Edinburgh, at the time. Although he was off school for some

2 As reported in the first school report card found for Eric's time there in 1928–29, when he was scoring marks in excess of 90 per cent in almost all subjects, a standard which he maintained through his primary years.

3 Eric kept the school cap badge amongst his most treasured possessions.

4 Better known in Britain as German measles.

months, he weathered the illness and emerged physically unaffected. Despite this, he very nearly died when he fell out of a first-floor window and impaled himself on railings below. As luck would have it, a passer-by saw the incident and called an ambulance, with Eric said to be very fortunate that his spleen had not been pierced. That would have been 'curtains', as he would say.

While Eric had his own health problems, his mother suddenly died after a long period battling tuberculosis. He was bereft. Euphemia had showered him with love and made him feel secure, and now she was gone.

Following his wife's death, and with his tailoring business suffering, Robert decided it was time for a fresh start, so with Eric he moved to Galashiels, in the Scottish borders. There he opened a 'sweetie shop', where he was known as 'Pa Brown'. This, of course, made him enormously popular, not just with Eric, but also with his friends.

When Eric finally started at the Royal High School in the spring of 1931, the change of address meant he faced a daily train commute of an hour each way. This could be gruelling, especially as Eric was still overcoming the after-effects of his illness and injuries, as well as the death of his mother. The boy who would become the world's greatest test pilot initially suffered from motion sickness on the train from Galashiels to Edinburgh.

Despite these challenges, Eric was rated 'very good'

at the school, where he rose through the ranks of the Boys' Brigade[5] until he reached non-commissioned officer rank, and also excelled at sports, particularly rugby and gymnastics. While Eric received top marks for English, Greek and History, his greatest passion was for modern languages. He grasped French very quickly, and in the 1932–33 school year German was also added to the school curriculum. He found he was a natural, much to the delight of his father.

Eric would later have a chance to visit Germany properly, bearing witness to the birth of the Nazis and attending one of the twentieth century's most notorious events in the process.

5 The Boys' Brigade has its roots firmly in Scotland and was created in 1883 in Glasgow. It is still active today, organized on quasi-military lines with a strong religious input, and retains its traditional values.

The great Nazi extravaganza of 1936 attracted visitors from across the globe. Eric claims to have been there in the stadium when the black American athlete Jesse Owens achieved gold, but there is precious little evidence to back up his story. This is one of the great unsolved myths of Eric Brown's extraordinary life.

2

Germany

Following the First World War, the 1919 Treaty of Versailles imposed a number of restrictions on Germany, with the aim of preventing any further global conflict. These restrictions included surrendering 25,000 square miles of territory, and seven million people along with it, as well as paying the equivalent today of US$269 billion in reparations. Military restrictions were also tough, as among other things the treaty prohibited Germany from building powered aircraft and weapons of war. It even went so far as to ban the training of aircrew. Unsurprisingly, conditions in Germany became volatile, and resentment started to build as other European nations apparently powered ahead. Soon after, a First World War veteran, who had been wounded in action, sought to change all of this.

Adolf Hitler never forgot or forgave the German surrender, nor the sanctions European nations imposed thereafter. Following a period in prison, after he had tried to seize power in a failed coup, he began to attract support by ferociously attacking the Treaty of Versailles, while promoting all things he believed to be Germanic, which in his mind included anti-Semitism. By 1932, such was his force and belief, in contrast to the struggling German economy, he helped to lead the far-right Nazi Party to power, although it did not yet have a majority. That all changed on 30 January 1933, when President Paul von Hindenburg appointed the increasingly popular Hitler as Chancellor, which saw him succeed Hindenburg when he died a year later.

With this, Hitler quickly manoeuvred his power to achieve full control of the legislative and executive branches of government, while also suppressing all remaining opposition in any ways he saw fit. Abolishing the office of President, and merging its powers with those of the Chancellor, Hitler became known as the *Führer und Reichskanzler* and turned Germany into a one-party dictatorship. Once Germany was fully under Hitler's control, he then turned his eyes to exacting revenge for the so-called injustices of the Treaty of Versailles.

By printing money, and seizing assets of his opponents, including all Jews, Hitler was able to finance an intense reconstruction and rearmament campaign, which saw unemployment fall from six million in 1932 to one million in 1936. On the surface, it appeared that Germany was respecting the Treaty of Versailles, while also curbing the growing resentment throughout the country by making it more prosperous. Most European nations thought this was a good thing, yet all was not what it seemed.

Twenty-five years before, Germany had dominated the world with its technology, particularly in aviation. Hitler wanted it to do so again, especially as his predecessors in Berlin had secretly laid much of the groundwork. Their plans for an air force had started in the late 1920s, with 'sports flying clubs' being set up, which appeared to concentrate on piloting gliders and pilot skills and thus were supposedly 'non-military'. However, turning a blind eye to the Treaty, the Soviet Union also allowed German pilots to undertake combat training, primarily because in the 1920s and 1930s it was still thought that aircraft were not able to become dominant weapons of war.

Hitler had immediately recognized the power of an elite German air force, and in 1933 he had secretly set up the Luftwaffe. By 1935, such was his power, and disdain for the Treaty of Versailles, that Hitler felt bold enough to make the Luftwaffe public, putting it under the direction of a flamboyant head, Hermann Göring, a First World War fighter ace. By 1936, almost miraculously, two dozen new aircraft designs from Willy Messerschmitt, Kurt Tank, Claudius Dornier, Ernst Heinkel and other talented men were being flown and tested. Many of the designs were even sent to the Condor Legion during the Spanish Civil War, where these aircraft were put into action for the first time. These details rarely made the press, especially in Germany, which wanted to avoid the watch of its anxious European neighbours.

It was against this background that, in 1935, Robert Brown decided to take his fifteen-year-old son Eric to Germany. The specific reasons for the visit are, however, unclear. Eric would have been learning German at the time in school and was hoping to study languages at university, so perhaps the visit was to help Eric become more fluent.

In any event, on 19 August 1935 Eric travelled to Germany on his father's passport, whose photograph shows Robert as a small man, with a small mouth, large ears, receding hair and an altogether undistinguished look. In contrast, the teenage Eric, while small in stature, boasted a fashionable centre-parting and cut a dash. There was naturally no physical resemblance.

Of their first visit to Germany little is known, other than that Robert's passport has stamps for passing across borders for trips to Aachen and Königswinter, where the academy of National Socialism, the Adolf-Hitler-Schule, was perched on the cliffs, high above the small town on the Rhine. The one memory Eric recalled of this trip was seeing a military procession march through the street. The black and red uniforms, the clicking heels and the pomp and ceremony all left their mark on him, especially in the medieval castle setting, against the sparkling, historically significant river below.

If for nothing else, the trip was memorable purely for this moment. With his own eyes, he had seen the Nazis in their infancy.

Eric claims in his autobiography that he and his father visited Germany again in 1936. This time the reasons for the trip are ostensibly clear. He contends that they were invited to attend the Olympic Games in Berlin, as well as a series of aviation events, which would debut a host of revolutionary flying machines. And yet, it is this trip, more than any other, that raises some tantalizing questions.

Although there is documentation, and a trail, that supports the Browns' 1935 visit, there is no documentation to support a visit the following year. Robert didn't have the required visa for this period, while there are no entry and exit stamps in his passport, which would have been essential in a totalitarian state like Germany. Tickets for the Olympic and aviation events were also regulated and exclusive. For Brits, it was strictly by invitation only. How, then, would a man like Robert Brown, a lowly sweet shop owner from Scotland, who had served as a labourer in the First World War, have obtained entrance?

We know that Robert held himself out to be a Royal Flying Corps officer, so perhaps he somehow used this fantasy to obtain the tickets. Another possible answer is that one of the many pro-German organizations, which were common in the 1930s, especially in Scotland, might have invited him. The Anglo-German Fellowship, formed in January 1935 and disbanded in December 1940, boasted many British society figures, even aristocrats, in its membership, as well as ordinary working men and women. Yet there is no record of Robert being affiliated to such a group, and this still doesn't explain his lack of visa or passport stamps.

Whatever the facts of the 1936 visit, Eric's autobiography describes how, as a sixteen-year-old, he found himself in Berlin for the Olympic Games. It was the most exciting place he had ever been. While the previous summer he had visited border towns like Aachen and Königswinter, this time around he was at the heart of the action. Berlin intoxicated him. Not just its bright city lights, cabaret clubs, vibrant street café-culture and excellent weather, but, after years of hurt, the city was in party mood for the Olympic Games. Red, white and black banners, emblazoned with the swastika emblem, hung from the lamp posts of every major thoroughfare, while futuristic Zeppelins flew in the sky, ready to quickly transport Olympic newsreel footage to other European cities. Eric would have been unaware, as were most other foreign visitors, that beforehand the Nazi regime had temporarily removed anti-Jewish signs and swept all gypsies off the streets and interned them in a camp at the edge of Berlin. Political opponents were also arrested and placed in the Sachsenhausen concentration camp.

Hitler was determined to use the games to show off Germany's technological developments and the superiority of its Aryan athletes. One American commentator wrote about the Olympic opening ceremony that it was 'almost [a] religious event, the crowd screaming, swaying in unison and begging for Hitler. There was something scary about it; his cult of personality'.[1] It was certainly an amazing spectacle. At the new 100,000-seater Olympic stadium, nothing had prepared Eric for the fly-past of the huge German airship *Hindenburg*, the release of 25,000 pigeons into the clear blue sky or the arrival of the Nazi Party leaders and their entourage. In later years, Eric would recount two particular memories. The first was the handshake between Hitler and the black American athlete Jesse Owens, after what Eric called an 'elegant demonstration of speed'. The second was his impression that Hitler's speeches were full of incomplete sentences, guttural sounds and bad syntax. The Royal High School's teaching of good German grammar had clearly made its mark on him.

Perhaps even more than the Olympics, Eric and Robert looked forward to attending the German aviation show. But before those events it appears that something quite incredible happened that would have a major

1 George Bebauer, *Hitler Youth to Church of England Priest* (Createspace, 2014).

impact on Eric's life. Somehow, they were invited into the inner circle of Wilmersdorf society by none other than the legendary Ernst Udet, the highest-scoring surviving fighter ace of the First World War.

Not only was he a war hero, but Udet was now renowned worldwide as a daring aerobatic pilot. His barnstorming display at the National Air Races at Cleveland, Ohio in 1931 had made the front pages of American newspapers, and now almost his every move was recorded in Germany and flashed around the world. While Udet was no ardent Nazi, the party still capitalized on his popularity, and he starred in the Nazi propaganda films *Hitlerjunge Quex* (Hitler Youth Quex) and *Wunder des Fliegers* (Wonder of Fliers). As regular visitors to flying displays and readers of magazines such as *The Aeroplane*, the Browns were well informed on current aviation news and would have certainly known who Udet was. Eric may have even seen Udet's flying exploits on the Pathé News newsreels during his trips to the cinema.

By 1936, Udet had just celebrated his fortieth birthday, but his receding hairline and shades of grey, coupled with a sallow complexion, made him look a decade older. His frequent drinking, smoking and late-night parties certainly would not have helped. He was also feeling the strain of his appointment with the newly revealed Luftwaffe – initially as the Inspector-General of fighters and dive-bombers, but now in a more central role in the Reich Ministry of Aviation – the Reichsluftfahrtministerium – in the centre of Berlin. He was one of the top three officials, and his responsibilities were taking their toll, as was the reduction in pay: an official's salary was far lower than the fees generated from stunt flying and movie making, with which he had funded his lavish lifestyle in the late 1920s and early 1930s.

During the Olympics, Udet held court in his large apartment at Pommersche Strasse 4. Eric remembers it was 'like the waiting room at Manchester Central Station', because people were coming and going all the time – most of them from the emergent Luftwaffe. In attendance were also noted inventors such as Karl Franke, who was then pioneering the first effective helicopter, which was being developed by Heinrich Focke and funded by Hermann Göring's Reichsluftfahrtministerium.

Another frequent visitor to Udet's apartment was 24-year-old Hanna Reitsch. At this stage Reitsch had only been a pilot for five years, but she was already a national aviation sensation, and, unlike Udet, she was a fanatical Nazi Party member. National and international newspapers eagerly recorded her record-breaking flights, and the Nazi Party revelled in her Aryan looks and manner. Although she is thought to have become one of Udet's lovers, she seems to have been more interested in flying than in men. In Eric's view, she 'batted for the other side, at least some of the time'.

Despite this, there was no doubting Reitsch's aerial achievements or her magnetic personality. Eric was particularly captivated by her. Reitsch was only eight years his senior, but this seemed insurmountable at the time. For now, Eric could do nothing more than admire this national icon from afar. However, in time their paths would cross again in very different, and dramatic, circumstances.

While meeting such celebrated aviators was a thrill in itself, things were soon to get even better. When Eric and Robert visited an air show at Halle, south of Berlin, Udet offered to take Eric flying in his Bücker Jungmann two-seat trainer. Eric could not hide his excitement. This really was a dream come true.

On the day of the flight, Udet took special care in ensuring that Eric was securely strapped into the front cockpit. Eric was puzzled as to why the German pilot was taking such trouble for a simple joyride, but all soon became clear. For the next thirty minutes, Eric's stomach was tested to the limits, as Udet thrilled the crowd, showing off a series of aerobatic feats. As the aeroplane was coming into land, Eric might have thought the showmanship was over, but Udet still had one trick left to play. At about 100 feet, Udet suddenly turned the Jungmann upside down and continued his approach until the top wings were just 50 feet above the ground, when he again flipped the biplane over the right way and 'flopped on to the runway'.

Rather than show any fear, Eric emerged from the cockpit grinning. It had been the flight of a lifetime. Udet was delighted with his reaction. He slapped him heartily on the back and said 'Hals- und Beinbruch', the traditional German fighter pilot greeting to a comrade who has achieved something great in the air, with the rather macabre meaning of 'break a neck and a leg'. In later years, after a long career in aviation, and having met many renowned pilots, Eric still claimed that Udet was the finest aerobatic pilot the world had ever seen.

On 4 August 1936, Eric and Robert watched Udet and Reitsch fly gliders over the Berlin-Staaken airfield in a series of demonstrations, with aerobatics and spot landings to show off the German lead in the sport. Staaken was a symbolic site, as it was not only the former Zeppelin factory, but also the airfield from which the first transatlantic passenger flights, in the Fw 200 Condor, would depart for New York. At the time, Germany had not revealed the fact that the Condor could be converted into a maritime bomber.[2]

2 The Focke-Wulf Fw 200 naming is fraught with confusion. 'Condor' was the official name, as all Fw types were named after birds, while Kurier was a relatively little-used name in Germany and much more commonly used by the Allies to distinguish the military type from the airliner. The common view seems to be that Kurier was purely an Allied name, but this was not entirely the case. It is indeed counter-intuitive, as 'Kurier' seems less militaristic than 'Condor', which is also sometimes, wrongly, identified as a Kondor.

It was a memorable summer and one that Eric would remember fondly for the rest of his life. As an adopted child, and as someone looking to feel like they belonged, he had been embraced by the world's greatest aviator and made to feel special. If he didn't think it before, then certainly from this moment on Eric always had a special place in his heart for Germany, not yet truly recognizing the emerging threat of the Nazi Party. Free from the shackles of Britain, where he hid his past, Eric felt he could be himself in Germany, just as the country was finding its own identity following the scars of the First World War. Together, they made a perfect match.

For now, Eric had to put such excitement behind him and knuckle down at school, where he was entering his final year, with an eye on attending university the year after. Growing into himself, Eric continued to excel, particularly at French, and became a star player in the school rugby team. Although slight, Eric was ferocious, flying into tackles and frequently taking down opposing players twice his size. While he had exceptional hand–eye coordination, he was also very vocal in urging his teammates on, his competitive instinct bubbling to the surface. Even the way Eric thought about the game marked him out as different from other boys. His diary records intricate details of matches, watched as well as played, all marked with the same fanatical attention to detail that would characterize his flying logbooks just a few years later. Such talents would all prove to be beneficial, and somewhat of a curse, as his life progressed.

When not a sportsman, Eric seems to have been a young man about town, with visits to the Playhouse theatre and the cinema, followed by stops afterwards at a favoured haunt, Crawford's Tea Rooms. He was also a member of the Dickson-Sandilands (1935) Club, the frequently attended Royal High School institution, where Eric liked to debate and champion key issues of the time, including 'This House does not believe that Mr Hore-Belisha[3] is dealing successfully with road traffic problems'. Again, Eric found he liked to talk, and especially liked to get his point across, if not always with tact and diplomacy in mind. He may well have been overcompensating, not only for his size, but also trying to prove he belonged in such company. Indeed, any challenge to a point of debate could earn an aggressive response. Yet while his interests were many and varied, aviation – and speed – remained his overriding passion.

In the summer of 1937, just as he heard he had been accepted to study Modern Languages at Edinburgh

3 It was Hore-Belisha who created the Belisha beacon and the 'zebra' crossing to aid pedestrians crossing busy streets, such a common feature of postwar British civic life. He was also Secretary of State for War from 1937 until Churchill became Prime Minister in May 1940 and was the architect of the Military Training Act in 1938, which called for more reservists.

University, with German as his primary subject, Eric came up with a ploy that would satisfy his need for adrenaline and might also pay for his university tuition. He convinced someone, perhaps an indulgent Robert, to buy him a Norton 500 motorbike, so that he could earn money riding the Wall of Death at Meadowbank Stadium, and then at RAF Turnhouse.

The Wall of Death was a very popular attraction, which had originated on Coney Island, New York. With his stage name 'Fearless Egbert', Eric's act involved carrying a fully grown male lion, riding either as pillion, or in a sidecar, whose amazing mane of golden hair streamed out behind him when at speed. Eric's act earned rave reviews, but unsurprisingly an inspector from the Royal Society for the Prevention of Cruelty to Animals stopped the performances. 'He didn't even ask the lion what he thought,' Eric would later say with a chuckle.

This was far from the only excitement Eric had in store that summer. Once more, Eric and Robert set off for Germany, this time obtaining visas from the German consul in Glasgow. Travelling across Europe by train, on this occasion their trip into the Third Reich was far less eventful. They spent most of their time sightseeing and walking in the Black Forest, the highlight of the trip being when they visited the Luftschiffbau Zeppelin, where Eric reported seeing a Zeppelin under construction.

By 1937, the Luftschiffbau Zeppelin company had been in the hands of pioneer airship pilot Hugo Eckener for twenty years, following the death, in March 1917, of Graf Ferdinand von Zeppelin, after whom the factory on the shores of the Bödensee was named. In September 1937, Eric may well have glimpsed the construction of the LZ 130, sister-ship to the ill-fated *Hindenburg* airship, which had been destroyed in a catastrophic fire in New Jersey as recently as 6 May. As a result, the company was struggling to keep the civil airship momentum going, which had seemed unstoppable just seven years before, when Hugo Eckener had captained a round-the-world Zeppelin flight, which had made the airship famous worldwide.

After returning to Britain on 7 October 1937, Eric enrolled at Edinburgh University and began his studies in Modern Languages. He seems to have split his time between his father's home at weekends for rugby and rowing and Edinburgh, where he had found lodgings close to the university. However, nothing matched his fascination with flying. So, in early 1938, when Ernst Udet invited Eric, and his father, to stay at his Pommersche Strasse apartment in Berlin, this was an invitation he could not refuse.

Once again, the apartment was full of renowned figures in German aviation. Such was the raucous atmosphere, at times things threatened to get out of hand.

During one of these infamous gatherings, the extrovert Udet introduced Eric to one of his favourite games. It involved one of his revolvers, usually a Colt Peacemaker, which he had brought back from America a few years earlier. The game required the shooter to use a mirror for vision, aim the weapon at a target behind the shooter's head and fire. It was a game Eric called 'hair-raising'. This might be somewhat of an understatement.

'He was an astonishing character,' Eric said of Udet years later. 'The fact that he invited me in to join the fun – which greatly improved my German and made me feel one of the group – was quite extraordinary.' After a few more drinks, usually more brandy, the shooting became even more wild and dangerous.

Udet would dress as the Flying Professor, his daring stunt-flying persona, and juggle bottles in front of the shooter to try and put him, or her, off their aim. Eric remembered that, despite this distraction, Hanna Reitsch always held her nerve and was a real sharpshooter. He said that this type of reckless behaviour was 'typical of the mood pervading Nazi Berlin'.

Rather than Reitsch's shooting skills, Eric most fondly recalled witnessing her fly the new Focke-Achgelis Fa 6, a twin-rotor helicopter, for its public debut, at the Deutschlandhalle in Berlin. The machine was the highlight of the Berlin Motor Show, and Reitsch showed a remarkable demonstration of controlled flying in what was still a revolutionary air vehicle. As we shall see later, this was to have a profound impact on Eric's life.

Another Udet devotee, whom Eric met at his apartment, was Werner Mölders, one of a pair of brothers who would become legends in the newly formed Luftwaffe. Eric remembered him as serious but always ready to talk about aeroplanes and air combat. This happily suited Eric. By early 1938, Mölders was about to go south to Spain, where civil war had erupted, and he had been posted to join the Condor Legion, which Germany had set up to support the Nationalist cause of Generalissimo Franco. Mölders suggested that Eric should get his pilot's licence and also go to Spain to fight as a volunteer. The money was good, and there was plenty of excitement for a young Scot with dreams of being a flying ace like his heroes. It seemed an exciting yet dangerous idea, something that certainly piqued Eric's interest. For now, he had to return to his studies in Edinburgh, but it seems that Mölders' suggestion was something he could not get out of his head.

Hidden away in one of Eric's logbooks is a notation about flying in Spain in 1938/39, at the end of the Spanish Civil War. In several of his detailed accounts of the handling characteristics of various aeroplanes, he goes into great detail about the Russian-built machines he flew there. Yet there is no documentary evidence to support he flew combat missions, and no passport records.

3

Flying High

By early 1938, Europe was spiralling into crisis. A few years before, the Italian leader, Benito Mussolini, had sent 100,000 troops to invade Abyssinia, while in 1936 the Spanish Civil War had erupted between Republicans, loyal to the left-leaning Popular Front government, and the Nationalists, a military junta led by General Francisco Franco and supported by Nazi Germany. Hitler was also beginning to dominate the headlines, particularly due to his increased anti-Semitism and more aggressive stance towards his European neighbours. War was beginning to seem inevitable, and across the continent there was now a real sense of urgency to get young men trained and eventually into uniform.

In Britain, HM Treasury loosened its purse strings, with the remit to teach as many young men as possible, with no previous aviation or military experience, to fly. When it was clear that the Service's Elementary and Reserve Flying Training Schools would be overwhelmed, the Treasury readily endorsed the request by the Air Ministry for funding to allow for training at civilian schools. In April 1938, the Civil Air Guard scheme was launched in conjunction with flying schools and clubs around the country, open to any fit person between eighteen and fifty. The cost was just 5 shillings an hour, and the Air Ministry even granted an annual tax-free bounty of £50, on renewal, to keep people interested and current as pilots. By December 1938, there were approximately 1,000 new licence holders from the scheme and a further 500 people, including forty women, under training. When war was declared the following year, these pilots would provide a very useful addition to the Fleet Air Arm, Royal Air Force and Air Transport Auxiliary.

Eric's summer in Germany had whetted his appetite to join his heroes in the sky as soon as possible. He posted his application to join the Civil Air Guard scheme and was overjoyed when it was approved. Almost immediately, on 15 August 1938, he drove to the Malling Aero Club, trading as Maidstone School of Flying, ready to fly.[1] There, he was taken under the wing of the Chief Flying Instructor, George Goodhew, who took him up in a DH 60G Gipsy Moth biplane trainer for his first thirty minutes of airtime.

In an intense whirlwind of activity, by the end of

1 From 1941, RAF West Malling operated Mosquito night fighters flown by such notables as John 'Cats' Eyes' Cunningham and Guy 'Dambuster' Gibson.

August Eric had logged nineteen flights, totalling about nine hours of airtime, and had also completed most of his ground schooling. As a result, on 11 September, Eric was deemed ready to undertake his first solo flight.

After twenty-five minutes of flying dual with Pilot Officer John F. Carroll, an RAFVR officer, who was later called up to the Royal Air Force during the Second World War, Eric was sent off alone. His first solo flight might have only lasted five minutes, but this was the moment that truly initiated his career as a pilot. There would, of course, be thousands of hours to follow, in far more dangerous circumstances.

At this early stage there was still much Eric wanted to learn. First and foremost, he still couldn't forget the thrill of watching Udet and Reitsch in Germany, performing a series of stunning aerobatic manoeuvres to the cheers of the crowd. With this in mind, Eric undertook several flights with the noted aviatrix Betty Sayer, a recently qualified Royal Aero Club 'A' Licence holder, instructor and, most importantly, a fellow aerobatics enthusiast. Sayer taught Eric a series of manoeuvres, and while they might have appeared to be nothing more than an opportunity to show off, many of these aerial tricks would later prove invaluable during Eric's combat flying.

After leaving West Malling, Eric set off looking for aerodromes where he could fly whatever aircraft were available to hire. He made the trip in his new, and much-prized, MG Magna L2 sports car. How Eric afforded such a stylish machine remains another mystery. He would say that he had a rich mill-owner grandfather who had paid for it. This is peculiar, as there was no rich grandfather. Whatever its origin, the MG certainly suited Eric's style. He was a smart young man and dressed fashionably, with a tendency to wear relaxed clothes, including Oxford bag trousers and open-neck sports shirts; he was also very good looking. Unsurprisingly, he also liked pushing the little sports car to its limits.[2]

In any event, Eric was desperate to fly whatever aircraft he could, and these included a Klemm L25, a forerunner of the Swallow at Croydon; the Comper Swift at Lee-on-Solent;[3] a Miles Hawk at Bekesbourne, home of the Kent Flying Club near Canterbury; and then at Southend a Zlin Akrobat, designed in Czechoslovakia, with a futuristic enclosed cockpit. It was a broad chorded-wing aeroplane, with very definite sailplane roots and a great aerobatic performance. A later service logbook for 1938 has a note, in Eric's unmistakable hand, of 193 hours

2 In 2012, he told his friend the aviation and motor-racing artist Michael Turner that his ambition was to get his first speeding ticket before his ninety-fifth birthday. He succeeded a few weeks later.

3 Technically an airfield and headquarters of RAF Coastal Command at the time and a place where Eric undertook catapult training in 1940.

of flying in Gloster Gauntlet biplane fighters, as well as the heavy Fairey Gordon biplane at Perth. However, the details surrounding this remain a mystery.

Flying a number of different aeroplanes, with their own flying characteristics, including the all-important 'numbers' – landing, stalling and speed not to exceed – was a crucial learning experience, particularly with war in Europe now looking inevitable. However, it was the civil war in Spain that first seemed to offer Eric a chance to fight in combat, and he wasn't too picky about which side he fought on.

The conversation with Mölders in Germany had first planted the idea in Eric's head, but it was another fortuitous encounter that gave him the final nudge. 'It all started with a man in a pub in Edinburgh,' Eric later told the filmmaker Nicholas Jones. 'I thought the man was recruiting for Franco – that would have been the Nationalists – but it was the Republicans.' This was, of course, on the opposite side of the Germans, who backed the Nationalists.

By late 1938, when it was clear that the Nationalists were going to win, the Republicans desperately offered many incentives to foreign pilots who volunteered to fight for them. They included flying and combat training, as well as being paid US$1,500 a month, with a bonus of US$1,000 for each aircraft brought down. In 1938,

US$1,000 would have been equivalent to £35,000 today. These cash figures were enough to attract many young men, like Eric, from across the world.

It appears that Eric decided to go to Spain during the Edinburgh University Christmas break. Much of what is said to have happened next remains clouded in mystery, and where Eric actually flew in this period is still open to conjecture. His logbooks are not always clear, but they suggest that he fought in combat over Figueras, near Barcelona, in which he claimed to have scored hits on a Condor Legion Messerschmitt Bf 109B, the latest German fighter which Göring had supplied to Franco's forces. He also claims to have shot down two Italian-built Fiat BR 20 medium bombers of Mussolini's Aviazione Legionaria, whilst flying the I-16bis. His records indicate one Fiat was shot down near Vilajuiga, near the French border, when it was probably attacking fleeing soldiers and civilians, and the second over Bayoles, probably in similar circumstances.

Contemporary accounts say that the Fiat bomber was sturdy and reliable, and was often able to evade the Republicans' Polikarpov fighters, so it says a lot for Eric's early skills as a fighter pilot if he was able to shoot down two of only a handful of Fiats lost to air combat or ground fire. In fact, if Eric's notes are to be believed, he would have accounted for 50 per cent of the Fiat casualties during the whole Spanish Civil War!

There is a reference in a later logbook where Eric refers to flying two Russian-designed fighters, the diminutive Polikarpov I-15 Type 5 and its monoplane successor, the I-16 Rata. The Rata was one of Eric's favourite aircraft. It was designed by Nikolai Polikarpov and Dmitri Grigorovich and was the world's first low-wing cantilever monoplane fighter, with retractable landing gear. He certainly found the Rata far more favourable than the Republicans' little biplane I-15b Chato ('snub-nose'), known as Chaika ('Seagull') in the Soviet Union. In his major work on the conflict, *The Battle for Spain*, Antony Beevor explains that the Chatos were manufactured at Sabadell but 'the 45 aircraft . . . produced in the last three months of 1938 did little to make [up the] losses' suffered in previous campaigns.

Eric's logbook contends that he flew the Russian-designed fighter between 23 December 1938 and an unknown date in the following month, for 61.5 hours and 67.3 hours. That's a lot of flying, even in wartime. These dates correspond to the Battle for Barcelona, which began on 23 December. As the Nationalist forces attacked on a wide front from the north, after some fierce fighting, Barcelona was captured by Franco's forces on 26 January 1939. Just three months later, it was all over, and Franco had won.

There are still plenty of unanswered questions surrounding this period. Did Eric fight for the Republicans or Nationalists, or both? Eric claimed, in a postwar newspaper interview and elsewhere, that in 1936 he became a member of the International Brigades, which were military units set up by the Communist International to assist the Popular Front government. This cannot be correct. At this time Eric would have been sixteen years old and still at school, with no recorded absences. The International Brigades had also folded in September 1938, so by the time Eric reached the required age, there was no need for more volunteers. Most damningly, there is no record of an Eric Brown on the official Scottish listing of volunteers, which appears to be a complete record of those who went to fight against Franco's Nationalists. So when exactly, and where, did his air combats take place? It seems we may never know the full truth. Even how he got to Spain, and out again, is uncertain. There is also a story that Robert Brown had to fly to Spain to spring Eric from jail, but there is no evidence, passport or otherwise, to show that this actually happened.

By his account, Eric returned to Britain in February 1939 and then continued university life, concentrating on his studies, where he was soon scheduled to spend a year in France as a teaching 'assistant', teaching English and improving his language skills. However, events were now quickly unfolding, and all of Eric's plans were set to be turned on their head.

While just a year previously, Prime Minister Neville

Chamberlain had declared 'Peace for our time' after Hitler had signed the Munich Agreement, the German Führer had reneged on it, seizing all of Czechoslovakia in March, with his eyes then set on invading Poland. And yet, just as things were heating up, Eric had plans to return to Germany.

Eric loved Germany – the people and the places he visited in the late 1930s. He was not a Nazi sympathizer despite the swastika on his MG's radiator grille, which was a precaution recommended to anyone travelling in 1939's Germany. This picture is taken near Augsburg in late August and shows a very relaxed Eric with a young lady looking on. In later years, Eric would show this photograph when discussing fast cars but was always coy about the lady's name. He was arrested and deported from Germany on 3 September but was allowed to keep the MG so he could drive back home to Britain.

4

The Great Escape

Throughout the summer of 1939, Hitler stepped up his demands on the Polish government in Warsaw, while also signing an alliance with Italy and a non-aggression pact with the Soviet Union. And yet on 6 July 1939 a seemingly untroubled Eric applied to go to France as a teaching assistant. Soon after, the French education authorities in Paris invited Eric to report to the Lycée de Jeunes Gens de Metz on 9 October 1939. However, an official Scottish Department letter revealed the wider context to his appointment:

> With reference to your application for a post as assistant in a French school or college during the session 1939–40, I am directed to state that the Department have [sic] been informed by the French authorities that they will be able to offer you such a post at the Lycée, Metz.
>
> In view, however, of the provisions of the Military Training Act, 1939, I am to ask that you will inform the Department whether you will be free to

> take up this appointment, i.e., whether you have been granted permission in terms of section 1(3) of the Act to postpone your liability to be called up for military training.

By this time in 1939, war with the increasingly aggressive Germany was becoming ever more likely, and just weeks later, on 12 August, the first military call-ups were made, when some 750,000 eligible young men received a telegram telling them to report to a variety of enlistment centres. Eric was not among them, as he had already registered as a student-teacher for 1939–40, which, as a reserved occupation, was not subject to immediate call-up.

This was indeed a relief to Eric, as before starting work in France, and despite the concerning rumblings of war, he intended to spend some time in Germany. He therefore visited the German Consulate in Glasgow to apply for a visa, and two days later, with it stamped into his passport, he set out from Galashiels in his MG sports car, with his father opting to join him for the first part of the journey.

A major conflict might have been on the horizon, but for now Eric chose to ignore any signs of war. He loved Germany, and its people, and like so many young people of the time he still thought that war was highly unlikely. He may have been encouraged in this belief by his father.

The stories of the dreadful carnage of the First World War played heavily on the minds of everyone in Britain, France and Germany. It was unthinkable for many that there would be another such war, especially with the constant barrage of pro-German propaganda. Many thought Winston Churchill was a 'war monger' and preferred to believe what Hitler's Propaganda Ministry told everyone: that Germany wanted peace and just needed to assert its rightful place in the world. Moreover, when war had seemed inevitable only a year earlier, the Munich Agreement had been signed, apparently stopping it in its tracks. Surely the powers that be would see sense once more and another Agreement would put a stop to this madness. That was at least the view of some, who clung to hope rather than analysing the facts.

After taking the early-morning car ferry from Dover to Dunkirk, Eric and Robert drove through eastern France to the border with Germany. There they found the first signs of tension, as French convoys drove eastwards, carrying less-than-eager-faced youngsters and grave-faced veterans of the previous conflict. Still, Eric nosed the MG past the Maginot Line, on which France staked so much, and then late on 1 August, with the Siegfried Line[1] clearly in sight, he drove through the border

1 More accurately, the 400-mile-long Westwall, planned in 1936 and under construction 1938–40.

queues close to Saarbrücken. In order for Eric to enter the country, he registered with the German authorities as a British student. This was something that would later prove vital.

Continuing on the new *autobahn* from Karlsruhe, with the canvas roof down, the little MG purred through the forests of the Swabian Jura, before Eric and Robert finally arrived in the recently reoccupied Rhineland. With blue skies, and harvests being gathered, it was all intoxicating, and war couldn't seem further away. Rather than push on to Lake Constance, Eric and Robert decided to linger in Ulm for almost a week.

The picturesque ancient city, situated on the Danube, had been central to so much of German history in the previous 500 years. Climbing to the top of the minster's roof, Eric and Robert could see the Alps, 60 miles to the south-east, while across the Danube was Bavaria, which was high on their 'to do' list. A hundred years beforehand, Bavaria had been its own kingdom, and less than two decades before, in Munich, had spawned the Nationalsozialistische Deutsche Arbeiterpartei – the Nazi Party – which had galvanized the new Germany. Signs of this were now clear from the roof, as everywhere Eric and Robert turned there were the signs and slogans of the new Reich.

On 7 August, after enjoying the week seeing the sights, and the local delights of beer and sausage, Eric exchanged more sterling for Reichsmarks, fuelled the MG and bade farewell to the city. Following a quick trip to neighbouring Switzerland, to deposit his father at a railway station, Eric returned to Germany, where he was scheduled to spend the rest of the month.

There had been no anti-British propaganda when Eric had visited Germany in early 1938, but now there were posters in railway stations on the north shore of Lake Constance depicting British soldiers destroying Arab villages, with the legend 'Gott Strafe England'.[2] For the first time Eric felt a little uneasy. He took the precaution of buying a swastika badge for the grille of the MG, which looked somewhat incongruous next to the Union flag. He also carried on his lapel a pin with the Nazi Party emblem. This in no way represented a political allegiance but displayed caution in these uncertain times. There would have been no point in standing out in a Bavaria that had almost totally embraced the new Führer and new National Socialist Workers' Party state.

Despite the area being in the grip of the Nazi Party, Eric found German hospitality as cordial as ever. However, while rationing had yet to be imposed, shortages were beginning to be felt as the government prepared for the inevitable. Nevertheless, Eric

2 Literally 'God punish England'.

was determined to enjoy himself. He was popular with the locals, who admired his self-confidence, and he even took one newly met friend to Switzerland, via the Lake Constance ferry, for a day trip.

Another trip took Eric to Friedrichshafen, where at the Löwental grass airfield he saw workers toiling feverishly. Claudius Dornier was testing amphibians (flying boats with the alternative of a wheeled undercarriage for land use) and Fliegeuder Bleistift or 'Flying Pencil' bombers for the Luftwaffe. By the time Eric saw them on production test flights, they were fully marked as warplanes, and Dornier was speeding their delivery for the forthcoming war, by then just days away. Also on the airfield was Zahnradfabrik Friedrichshafen (better known by its initials ZF), making ball-bearings for almost everything in Hitler's war machine. The late Ferdinand von Zeppelin's firm Luftschiffbau Zeppelin was also still building giant airship designs and supporting Dornier-Werke and Maybach Motorenbau's tank engine plant. For Eric, the attraction was, however, not industrial but purely practical – the *Sportsflughafen* (now Friedrichshafen airport) was where gliding took place every weekend, and he was there to observe, not quite realizing that many of those participating were practising their flying skills for conflict.

The summer was turning into everything Eric had

dreamed of, but then it all came crashing down. On 24 August the British Embassy in Berlin issued an order for British subjects to leave Germany immediately. Even then, Eric neglected to leave. Either he was blissfully unaware of the warnings or he chose to ignore them. The German authorities were, however, very much aware of the students and other visitors in the country and were now keeping tabs on them.

On 1 September, while Eric planned a weekend motor-touring to Munich, 500 miles to the north and east, German troops crossed into Poland from East Prussia and Slovakia. Unaware of this, Eric packed the MG and drove through the harvest fields and ripening vines to the ancient city of Augsburg, which had once been the seat of the Holy Roman Empire's bishops and is still the gateway to Bavaria.

It seems that the passenger seat was occupied by a young lady. It must have been idyllic, travelling through a country he loved, in his cherished sports car with the roof down in the late-summer sunshine, with a mysterious female for company. Yet when he and his companion stopped in a *Gasthaus* for the night, Eric would not be able to ignore the war for much longer.

There are several of accounts of what happened next, but what we know for certain is that on that Sunday morning, 3 September 1939, two uniformed members of the German state security service, the SD,[3] arrived at the *Gasthaus* and ushered the landlady upstairs to wake the *Britisher* student and his companion. Eric remembered, 'I wasn't in a position to argue, I was taken away, everything in my room was confiscated, and my car was also taken to the SS [barracks].'[4]

Eric claimed it was morning when the landlady knocked on the bedroom door, but if this was the case it must have been late morning, as war between Britain and Germany was not declared until 11 a.m. London time, which was 12 noon Berlin time. From that point onwards, all Brits in Germany were persons of interest to be rounded up. Eric would later claim that the SD had jumped the gun by six hours, but this seems unlikely, as the British government's ultimatum on Poland came as a surprise to the German government when it was handed over in the early hours by Sir Neville Henderson, the British Ambassador in Berlin, to Joachim Ribbentrop, the German foreign minister.

Held in a police cell, Eric was repeatedly quizzed about his intentions and reasons for being in Germany. He lied and told them he had a 'teaching job at Salem'. The fact that his official documents marked him out as a teaching assistant certainly helped embroider the lie,

3 The Sicherheitsdienst (SD) was the intelligence wing of the Schutzstaffel (SS).

4 Eric later recalled this in his autobiography, *Wings on My Sleeve.*

as did the fact he had travelled in Germany before, that he spoke the language and that he could talk about the sights of Berlin – not to mention his contacts with the senior members of the new order in Germany's capital. All of this worked in his favour, and yet he was still held captive as his guesthouse room was searched for any incriminating evidence. This must have all been tremendously frightening for Eric, but thankfully the wheels of diplomacy were already in action.

As there were German students in Britain, reciprocal exchange arrangements were being negotiated, courtesy of the Red Cross. The humanitarian organization, which would play such an important role in the Second World War, subsequently arranged for over twenty British students who had been in Germany to be set free in Switzerland. To Eric's relief, after spending three days in a cell, he learned he was to be one of them. Had he been interned, he might so easily have missed the whole period of conflict, as happened to many others.

Sandwiched between two 'burly' SD officers, Eric was taken to the Swiss border in a Mercedes staff car. As they came to a stop, Eric thought he was going to be expelled with only the clothes on his back, with his belongings and cherished car left behind. However, upon looking over his shoulder, through the small rear window, he was surprised to see a large, uniformed sergeant driving his little Magna sports car. Eric recalled that the sergeant was so tall that his head projected above the windscreen, and that he was so broad that he hardly fitted into the diminutive British machine. 'A bit of a squeeze for a man who was obviously fond of his beer and sausage,' he would later say.

To his amazement, Eric was told to take his car. 'You've taken my books, clothes and money,' he said, 'so why not my car?' The SD officer's reply was almost contemptuous, 'Because we have no spares for such a machine.' Eric was overjoyed, not only at having his cherished car returned to him but also at the prospect of being able to drive back to Britain. Still, his troubles were far from over.

In September 1939, crossing into Switzerland by the land border with what had been Austria, but was now part of Greater Germany, should have been a relief. After all, the SD had assured him that he was expected and would have little difficulty in passing the border controls in neutral Switzerland. In the event, when Eric presented himself at the customs checkpoint, the Swiss *gendarmes* who greeted him were 'stony faced'. For the next few hours, Eric and his car were detained by the Swiss *gendarmerie*, as he was once more locked up in a single tiny cell, with no water and very little air. What should have been a relief began to seem a continuing nightmare, as he was treated less favourably than by the SD. After some frantic telephone calls, the British Embassy in Berne stood surety for him, until consular officials could telephone through a confirmation of his *bona fides*.

Without receiving so much as an apology, Eric was finally tossed out on to the street. Not wishing to hang around, he quickly departed for Berne, the Swiss capital. At least he had his car and some of his belongings – such is German efficiency – but alas, no money.

When he arrived in the Elfenau diplomatic district of Berne, His Majesty's Ambassador, Sir George Warner, settled Eric in a chair in his study and, along with another man, probed him on what had transpired in Germany. Although Eric couldn't be sure about the identity of the other man, who was taking notes, he was probably the Secret Intelligence Service's station chief in Switzerland. It did not take long for Sir George to understand that Eric wanted to return to Edinburgh as soon as possible and join the war effort. He subsequently passed Eric to Sir Frank Nelson, his senior consular official in Basel. Following yet another debrief, Nelson arranged accommodation that night in Switzerland. He also gave the exhausted Eric petrol coupons to speed his journey through France to Dieppe, where he could catch the Newhaven ferry.

As a tide of Brits fled across Europe, in the hope of returning home, there were yet more issues facing Eric at Dieppe. Along with the British Embassy staff from Berlin, and many others caught up in the sudden events in Europe, Eric experienced difficulties getting personal effects, including his car, back to England. After engaging in protracted negotiations with the ferry company, Eric discovered that only military and diplomatic vehicles could be carried, as war emergency rules had been invoked.

Just as Eric became resigned to having to leave his MG behind, he finally had a stroke of luck. A Royal Automobile Club representative understood Eric's predicament and promised to find a way of getting the MG across the Channel. But for now, he would have to go as a foot passenger to Folkestone and take the train to London, and thence to Edinburgh. 'To my utter surprise,' Eric recounted sixty years later, 'the RAC got it back a month later, so I joined [the RAC] and I have been with them ever since.' He proudly wore the RAC badge on his car until he stopped driving in October 2015.

On his return to Edinburgh, Eric could no longer hide from the obvious. Britain was at war with Germany, a country and people he loved almost as much as his own, and he was going to have to fight them. This would be more dangerous and traumatic than he could have ever imagined.

PART 2

War Hero 1940–42

Eric Brown joined the Royal Navy in 1939, interrupting his language studies at Edinburgh University, and put on the uniform of a Naval Seaman Second Class. He duly reported for training at Gosport and was selected as a potential pilot. The rating on the left is his great friend Graham Fletcher, with whom he shared many adventures and several disciplinary hearings.

5

The Royal Navy

With a love for aviation, and war breaking out, there appeared to be only one place for Eric: the Royal Air Force. Yet when he duly reported to the recruiting office in Edinburgh, he was told it was currently full, and there would be a three months' wait. Eric became impatient. He wanted to get into the action, and the wait was more than a young, fit and enthusiastic nineteen-year-old could bear. However, he soon saw a note that would change the course of his life.

He learned that there was a shortage of pilots for the Royal Navy, which had been neglected in the pre-war build-up, and there was therefore an urgent call for volunteers. It was here that Eric decided that a life in the Royal Navy, rather than the Royal Air Force, was for him, for no other reason than 'it would get me into the air right away'. But it was a very different prospect to the Royal Air Force.

While the Royal Navy had once been consigned to ruling the high seas, in 1909 it began to experiment with the new-found benefits of aviation, by constructing an airship and then a warship from which planes could fly when at sea. In 1911, its first pilots graduated, and a few years later, in 1914, the so-called Royal Naval Air Service (RNAS) provided fleet reconnaissance during the First World War. Its duties included patrolling coasts for enemy ships and submarines, attacking enemy coastal territory and defending Britain from enemy air raids, along with deployment along the Western Front and overseas.

In April 1918, following the end of the war, naval and army aviation forces were combined to form the Royal Air Force (RAF), with the Fleet Air Arm of the RAF formed on 1 April 1924. This focused on RAF units that would embark on aircraft carriers and fighting ships, with HMS *Hermes* being the world's first ship to be specifically designed and built as an aircraft carrier. However, with the Second World War just three months away, and it being recognized that in the intervening years the Fleet Air Arm had fallen significantly behind the Germans, on 24 May 1939 it was returned to Admiralty control and renamed the Air Branch of the Royal Navy. This was what Eric now hoped to join.

In early November, Eric, in civilian clothes, made his way to the Admiralty in Whitehall to attend an interview, which even in wartime was regarded as necessary to become a member of the Air Branch of Royal Naval Volunteer Reserve. As chance would have it, while he travelled south on the train, he noticed that the next compartment had armed guards. Looking more closely, he saw that behind the guards were two Luftwaffe air-crew, seated next to Royal Air Force Police minders. Having just returned from Germany, and with his German fresh in his mind, Eric sought permission to speak to them.

With the guards watching on, Eric discovered that the two airmen were survivors from a crew of five who had been flying in a Heinkel, which had the dubious place in history of being the first enemy aircraft to have been shot down on British soil since the First World War. The newspapers of the day had made a big play of the air action over the Firth of Forth on 28 October and of the skill of Flying Officer Archie McKellar, who was credited as the victor, with the Heinkel crashing at Humbie in East Lothian, near Dalkeith.

In conversation with the Germans, Eric discovered the crew were from the staff flight of Kampfgeschwader 26, known as the Vestigium Leonis unit, usually translated as the 'Lions'. This unit played a major part in the early air operations against Britain, and especially shipping in the North Sea. The survivors were the bomber's pilot, Unteroffizier Kurt Lehmruhl, and Leutnant Rolf Niehoff, the navigator-leader of the entire unit. To say they were unimpressed to be among the first Luftwaffe prisoners of war was an understatement.

Speaking in German, Eric tried to lift their spirits by assuring them that they would be well treated. They, in turn, thought that the war would at least be over 'by Christmas, one way or another'. If that was to be the case, Eric still wanted as much of the action as was on offer.

However, following a successful interview in White-hall, Eric found becoming a pilot in the Fleet Air Arm took more time than he had hoped. He was eager to get started right away but despite informing anyone who would listen that he already had 125 hours' flying time, it didn't matter. To his frustration, the Royal Navy insisted that all new recruits, whether they could already fly or not, should undergo recruit training as a Naval Seaman Second Class at a 'stone frigate', or shore estab-lishment, beside Portsmouth Harbour, called HMS *St Vincent*.

Disgruntled and frustrated, on 4 December 1939, Eric duly presented himself at the gate of the naval bar-racks, having taken the Gosport ferry across the har-bour. The naval training programme had changed little since the establishment had been created to train, first boy sailors and then, in the current 'emergency', poten-tial pilots and observers for the Air Branch.

Before Eric could think about getting into an aeroplane, he first had to learn basic naval skills, such as Morse code transmission, lights, flags, naval tradition, drill and basic marksmanship. All of this was overseen by the infamous Chief Petty Officer Wilmot, who was loved, feared, respected and obeyed simultaneously. Small of stature, but large of voice, he garnered instant respect and took no prisoners. Should a recruit not toe the line, Wilmot delighted in issuing out punishments, which included lighting the coal fires in the petty officers' office, cleaning the heads (toilets) and running around the parade ground until the recruit dropped or Wilmot finally took mercy.

The course wasn't easy, yet Eric passed out with the assessment of 'above average ability; good progress made; keen', earning 524 marks out of a possible 650, making him eighth out of 55 ratings. Now, at last, he could get down to business and earn his wings. Being the 'suicidal type', Eric hoped to be a fighter pilot, which would require protecting British convoys at sea, taking off and landing on an aircraft carrier and engaging with German aircraft. This certainly seemed to promise the adrenaline rush Eric was after.

After a short leave at home in Galashiels over Christmas, when Eric returned to HMS *St Vincent*, the postings were placed on the main noticeboard. Eric approached with bated breath. This could very well determine how he spent his time for the rest of the war. As Eric scanned the list of postings, he saw that he had been selected to train as a fighter pilot in 'the naval way' at HMS *Gadwall*, at Sydenham on Belfast Island. He was thrilled and couldn't wait to get started.

On 15 January 1940, an excited Eric arrived in Northern Ireland and found that he had been promoted to Acting Rating Pilot, effective that day. Such was his rapid progress that within weeks he was an Acting Leading Airman in the Royal Naval Volunteer Reserve (A) – the A designating his aviation specialist duties. However, as he edged ever closer to first-hand aerial combat, Northern Ireland was about to throw him a curve ball he didn't see coming.

Eric's first solo flight in a military aeroplane was in a Miles Magister. It gave him a lifelong admiration for the Reading-based aeroplane manufacturer, and a love of single-engined monoplanes. He was grounded for two weeks for unauthorized aerobatics and nearly died in a Magister showing off to his new girlfriend, Lynn, in Belfast.

6

Love and War

'Look, that's my boyfriend in that aeroplane,' shrieked the petite, bubbly schoolgirl, her hair in ringlets, pointing at the sky. Seconds later, the windows of her ground-floor classroom at Methodist College Belfast rattled at the sound of a Miles Magister training aeroplane roaring past, with Eric in the open cockpit.

A few days earlier, Eric had met the young girl while she was standing in the doorway of her family home in Elaine Street, Belfast. Her name was Evelyn Jean Margaret Macrory, known to all as 'Lynn', the seventeen-year-old daughter of local sports journalist Robert Macrory, stalwart of local civic society and the Presbyterian Church. Eric was instantly smitten. It was clear that this was 'the real thing'. In later life, Lynn would say that the same thought hit her: the twinkle in the eye and the naval uniform were enough, even before she spoke to the smooth-talking young Scot.

Having just met this beautiful young girl, Eric was keen to impress her, hence his unauthorized fly-by outside her classroom. He no doubt thought he was like his hero Ernst Udet, performing gravity-defying feats to the cheers of the crowd. Yet, soon after, it almost ended in disaster.

In a thick mist, he had been concentrating on keeping the Magister in balance, under control and close to the ground, straight and level, while also waving at Lynn. Suddenly, through the mist ahead, the school rugby posts flashed into view. He instantly turned to starboard, but the Magister didn't respond as quickly as he hoped. Clearing the rugby posts by just inches, Eric breathed a sigh of relief, but then out of the gloom, looming large, were the twin spires of St Peter's Catholic cathedral.[1] Collision seemed inevitable.

Desperately trying to pull the Magister clear, Eric closed his eyes as the port wing clipped the cathedral. With a loud bang, he watched as two feet of the mainplane, together with the aileron, fell to earth. Somehow, the Magister stayed airborne. He was extremely fortunate, as, although damaged, he was still able to land at nearby HMS *Gadwall*. However, as a result of that unauthorized flight, Eric was grounded for a fortnight by the Commandant of 24 Elementary Flying Training School, and it was recorded in his naval records.

Although not ideal, the two-week grounding for the

1 Eric would later claim that the cathedral had six inches knocked off one of the spires, but examination of the records does not match that account.

cathedral incident actually proved a real bonus. With time to kill he could spend more time with Lynn, who was impressed by his aerobatics outside her school. At their favourite rendezvous, the Crawfordsburn Inn, romance blossomed, and they talked of their hopes and dreams – Eric to be a pilot, and Lynn to be a singer. They might have seemed an unusual match, as Eric was more reserved, other than when he was flying, while Lynn was vivacious and outgoing. But deep down both loved performing for an audience. They wanted to be seen, and heard, and in time they would help each other to achieve just this.

While Eric furthered his romance with Lynn, he also used his suspension as an opportunity to study for the theory part of his course, which he passed top of the class, with an assessment as being 'above average/ exceptional'. Surprisingly, only his navigation skills were not rated perfect.

When his two-week suspension came to an end, and with a dire shortage of aircrews in the Fleet Air Arm, Eric was told he was being sent to RAF Netheravon in Wiltshire for advanced fighter training. This, of course, meant he would be posted away from Northern Ireland, and away from Lynn. They promised to write to each other and were determined that their romance would continue, but with war now erupting, and Eric soon to be posted overseas, who knew what the future held?

The course at Netheravon began on 21 April 1940, and it offered a new set of aircraft for Eric to master, including the Hawker Hart biplane bomber[2] and the Fairey Battle, as well as a series of air exercises, including navigation, blind flying and air combat training. All of the students were being watched to see who would make the grade. Using the large grass airfield at Netheravon, and that of neighbouring RAF Old Sarum, just north of Salisbury, Eric flew at every opportunity, desperate to prove his ability.

Time was now of the essence, especially when in July the Battle of Britain commenced, with the Luftwaffe raining bombs down on strategic targets and cities in an attempt to strangle Britain and force it to surrender. It didn't take long for Eric to get a glimpse of what this was all about.

On 21 July 1940, Lieutenant Robert Churchill, an RNVR Air Branch student, and his Royal Air Force instructor, Flight Lieutenant John Wray, were flying a Fairey Battle trainer near Stockbridge, on the return leg of a navigation exercise from Netheravon. This single-engined aeroplane was designed as a bomber, but even though it was powered by the same Rolls-Royce Merlin engine as a Spitfire, it proved slow and vulnerable when

2 A pre-war open-cockpit biplane which had been designed as a bomber for the North-west Frontier of India. Now, in 1940, it was being used to train the future fighter pilots of the Royal Navy and the Royal Air Force.

flown on operations in France in May 1940. Now a tandem-seat[3] navigation trainer, it was, however, perfect for this role as it allowed the student pilots an enclosed cockpit, and an appreciation of flying a single-seater.

As part of his training, Churchill was blind flying, with an opaque hood covering his part of the cockpit, so he had to fly solely by instruments. Suddenly, Wray heard a humming noise approaching. Glancing behind, his eyes widened as he saw a single marauding German Messerschmitt Bf 110 twin-engined fighter coming at them hard and fast.

Using all of his flying skills, Wray climbed vertically to evade the Messerschmitt, as it chased them across Hampshire and into Wiltshire. He almost succeeded in getting back to Netheravon by flying through blanketing cloud, but the Battle's Merlin engine cut out. With little choice, Wray had to put the aeroplane down at Shrewton, to the west of Netheravon, where the two escaped, shaken but unharmed.

The Messerschmitt, flown by Oberleutnant Friedrich-Karl Runde, with Feldwebel Willi Baden, continued on its journey to Old Sarum, another Salisbury Plain airfield to the south-east, where it found Acting Leading Airman John Seed in an unarmed Hawker Hart biplane. Runde

3 The student pilot sat in front of the instructor in a Perspex-enclosed cockpit.

brought the heavy twin-engined fighter into formation, while Baden opened fire with his single 7.92 mm MG15 machine gun mounted in the rear cockpit. The Hart was peppered with gunshots and erupted into flames. The wounded Seed managed to escape from the aeroplane, but his parachute did not open at such a low altitude. By the time students from the airfield arrived to Seed's rescue, he was already dead. Eric had seen the whole episode unfold to the east of Old Sarum aerodrome. It was a dreadful experience, but typically he learned something from it: 'A fighter pilot has to have a swivel neck.' [4]

With the perils of war now very much on his mind, Eric continued to learn to fly a wide variety of aircraft. According to his logbook, these included the Fairey Battle, the Hart, the single-seat fighter version of the Hawker biplane stable, the Fury Mk I and the Mk II. He also had the opportunity to fly the Hawker Hector army cooperation biplane at Abbotsinch, although exactly under what auspices is unclear. In doing all this, Eric displayed a wonderful capacity to jump from one aircraft to

4 Runde and Baden did not make it back to France that afternoon and were shot down by Hurricanes from No. 238 Squadron from nearby RAF Middle Wallop, which chased the Messerschmitt southwards towards the coast. Both aircrew survived the engagement off the Isle of Wight and the subsequent crash landing to spend the rest of the war in captivity in England.

the next. Likening it to riding a bike, he would later say, 'Once a type learned, never forgotten.'

However, just learning to fly a variety of aircraft was not enough. The Netheravon course also included blind flying approaches, which were extremely difficult. Under the hood of a Hart training biplane, Eric learned to fly solely by instruments. In five hours of Lorenz blind-flying,[5] he also used radio navigation to aid ground training on how to use glide slopes to airfields, and radio beacons, all under the careful tutorage of Flight Lieutenant Vaughan, fondly remembered by Eric. Such was his prowess, Eric's service record shows that he was being watched closely by the instructors. He was described as a 'quiet type but really has a strong and interesting character. Will make a good officer.'

On 25 August 1940, Eric finally finished his course at Netheravon, scoring 79.2 per cent, with 166 hours logged on the Miles Magister, Fairey Battle and Hawker Hart. He came out top of the class. It was clear he was destined to be a fighter pilot. Eric might have been excited by this, but at this stage in the war fighter pilots were dying at an average rate of 385 per month. Soon he would see why.

5 Lorenz blind-approach system for bad-weather flying.

An unexpected journey to Norway – Eric is seen here flying a Blackburn Skua dive bomber through a Norwegian fjord, chased by a Messerschmitt Bf 109 fighter. As Eric's first naval aviation logbook was lost in December 1941, the details of the incident have been pieced together from British and Norwegian records. As Michael Turner's exquisite art shows, it was a close-run thing.

7

A Taste of War

On 26 August 1940, Eric was commissioned and promoted to Probationary Temporary Sub-Lieutenant (A). With true British administrative zeal, his personal file was marked 'Ineligible for Prize Money'.[1]

Subsequently posted to an operational unit, 802 Naval Air Squadron, at RNAS Donibristle, Eric found a kindred spirit in Temporary Sub-Lieutenant Graham Fletcher, known of course as 'Fletch'. He was younger than Eric, and had joined the Fleet Air Arm straight from St Paul's School in London, but the two had plenty in common. Like Eric, he had already earned his pilot's licence with a civilian club, and like Eric he was also keen on adventure. Their partnership thrived on adrenaline, and they would volunteer for anything which involved flying. They would soon get a very unexpected, but dangerous opportunity.

On 10 September 1940, two Skua dive bombers were

1 If a ship was captured during war at sea, the ship's company which carried out the action would share in its value, pro-rata for rank.

flown north by Eric and Fletch, to RNAS Hatston on Orkney, where they were then to take part in a planned raid against oil storage tanks at Skålevik. However, on the Skuas' arrival, the commanding officer of 801 Naval Air Squadron, Lieutenant-Commander Ian Sarel, found that the two pilots who had been designated for the operation had fallen ill. He therefore 'invited' Eric and Fletch to take their place. 'Fletch and I were not dismayed by this turn of events,' Eric recalled, 'but indeed were rather pumped up at the prospect of some real action.' This was to say the least.

Subsequently allocated a Skua each, they were then both teamed with an experienced naval officer, called an observer, for navigation. The observer also manned the radio and operated the rear-facing Vickers K-type machine gun, which supplemented the pilot's four forward-facing Browning weapons of the same 0.303 inch calibre. In this case, Eric's observer was Lieutenant William Iliffe.

For the raid on Skålevik, Eric was to be in the second wave, led by Lieutenant (A) T. E. Gray DSC. Later in life,[2] Eric recalled the approach:

2 In conversation with the artist Michael Turner in preparation for the painting featured in Eric's last book, *Too Close for Comfort: One Man's Close Encounters of the Terminal Kind.*

The next day [13 September] had favourable weather, so we were off early on the long haul in weak mixture (to use the least amount of fuel possible) over to Norway in a gaggle of twelve. We two new boys carried a squadron observer and a 500 lb bomb; and it was indeed a long haul, made less tedious by having a companion in the rear seat. The transit was made at 1,500 feet until we sighted the Norwegian coast and began to climb to 7,000 feet, which we reached without opposition. We all lined up in the order briefed, and I found it absolutely exhilarating to hurtle down on the target, release my bomb and then push down, heading for the exit fjord.

After releasing his single bomb, Eric pushed down to sea level and made for the open sea. Going north along Hjeltefjorden, the retreating Skuas suddenly encountered defending Messerschmitt fighters. 'The whole scenario changed as a swarm of Bf 109E fighters pounced on us,' Eric remembered. While the Skua was an effective bomber, it had previously been found wanting against fighters. It was slow in comparison and required an experienced pilot if it was to out-manoeuvre a fighter in the class of a Messerschmitt. Despite intensive training, this was also Eric's first time face to face with the enemy, and a superior one at that.

Hugging the east wall of the fjord, Eric hoped to

restrict his 'opponent to a difficult stern attack,' giving his 'observer an easy target for his single machine gun'.[3] When Iliffe called out that a fighter was closing from astern, Eric reacted quickly. Using the relatively new fighter pilot dodge, he opened his dive flaps and slowed the Skua right down. This manoeuvre forced the chasing Messerschmitt to take avoiding action, spoiling his aim and making him climb up and have another go.

Despite this manoeuvre, some of the cannon and machine-gun fire from the German fighter hit Eric's Skua, wounding him in the arm, while also hitting Iliffe in the rear of the cockpit. Luckily, the latter had dropped down, so missed a direct hit, yet he was still concussed and in no fit state to continue. Using the skills he had picked up practising 'fighter evasion training', as well as some aerobatics, the wounded Eric somehow managed to evade the Messerschmitt fighters and returned to the Orkneys, not quite sure how badly wounded Iliffe was.

'It was a very long haul back across the North Sea,' he remembered, 'and although my arm wound ached, it was not bothering me as much as my concern for my Observer, and the fact that I could not see any of the other Skuas.'

On landing, while Eric and Iliffe received medical

3 *Wings on My Sleeve.*

attention, he learned that two Skuas had been lost near Bergen, and a third had ditched into the sea on the way back. To his relief, he found that Fletch had at least returned unharmed.

Soon after, Eric and his messmates were sent to HMS *Raven*, the 'stone frigate' name for the Royal Naval Air Station at Eastleigh. Located near Southampton, the airfield was also home to the aircraft manufacturer of the Supermarine and the Spitfire. At this time, the Royal Navy's Fighter Course 10A, at 759 Naval Air Squadron, was in a hurry to turn out new fighter pilots for the Fleet as quickly as possible. Eric did not disappoint. His formal report on 6 October 1940 records in typical understated English manner that he 'is very keen and quite good all round'.

Matters were, however, not always straightforward. On 12 November, while Eric was flying a Hurricane, it suffered an in-flight fire, and after crash-landing at Eastleigh, his face was badly scarred. He required urgent treatment, but after a few days in hospital recovering, Eric had had enough. He subsequently checked himself out and was determined to return to flying. Surprised that he was back so quickly, the Admiralty posted him to the newly built Royal Naval Air Station at Yeovilton, to where 759 had also decamped.

Located on the Somerset Levels, the air station which was to become HMS *Heron* was seen as the home of naval flying then, as well as today. With the Battle of Britain over, but German bombers still in evidence, local air defence in Somerset required bolstering, and the trained, if not experienced, pilots from the Fighter School were assigned to defend the new home base from German bombers. This was an area of huge importance, as the airfield also provided protection for the Westland Aircraft factory at Yeovil, as well as the aero-engine production facilities of Rolls-Royce and Bristol at Filton, all vital to the war effort.

With this in mind, on 19 November a pair of Heinkel He 111 bombers were spotted near Dorchester, making a direct track for Bristol. Eric, in company with his messmate Fletch, moved to engage the small formation, each choosing a bomber as their target. As Eric readied to fire, his target suddenly exploded, having been hit in the bomb bay by anti-aircraft fire from below. The resulting explosion caused a pressure wave, which made Eric and his Sea Gladiator biplane 'buck wildly'. On seeing the explosion, the second Heinkel turned away for France, jettisoning its bombs harmlessly in a field, but was forced to crash-land before reaching the coast. Both Eric and Fletch's Sea Gladiators had been hit by return fire from the German bombers, and as Eric put it, 'our German instructors made our graduation as naval fighter pilots real'.

Soon after, Eric was again called into action in a Sea

Gladiator biplane, with Sub-Lieutenant Hutchinson as his wingman. As they flew over the Shaftesbury area at 12,000 feet, the cry of 'Tally ho!' went out, as up ahead a formation of Messerschmitt Bf 110 twin-engined fighters came into view. 'We waded into a monumental dogfight,' Eric recalled, 'complete with [the Bf 110s forming] a defensive circle and other weird and wonderful tactics.'[4] Eric fired 'a short, sharp burst' of machine-gun fire from the beam 'at full deflection' to 'no apparent effect', with the enemy aircraft escaping into the cloud.

Although he had seen combat during 1940, Eric's efforts were never officially recognized, because he was not part of a designated squadron, and instead formed local air defence flights. Due to this, it was deemed that he, and others, would not be eligible for the Battle of Britain clasp or membership of the Battle of Britain Fighter Association. This was always a sore point for Eric. He would later rib the ace of aces, Johnnie Johnson, about his clasp for just one day on a squadron and no combat missions, whilst he had two missions, but not with an officially recognized squadron, so no Battle of Britain clasp. The truth is that the Battle of Britain had

4 In one version of the story, Eric gives October 1940 as the date. A later reference in his notes records this date as March 1941 and his aircraft being a Sea Hurricane Mk Ib. This seems unlikely, as the number and size of German formations in daylight had drastically reduced.

officially ended on 31 October, so Eric was timed out for a clasp anyway.

Following this, on 21 November 1940, Eric and Fletch were posted back to 802 Naval Air Squadron. The unit was based in Scotland but was rather nomadic, using both RNAS Machrihanish (HMS *Landrail*) on the west coast and RNAS Donibristle (HMS *Merlin*) on the Firth of Forth for aircraft carrier training. Ever restless, and eager to push the boundaries, Eric soon found himself in serious trouble.

Eric and Fletch were instructed to deliver Swordfish torpedo-bomber biplanes from RAF Sherburn-in-Elmet aircraft depot in Yorkshire to RNAS Donibristle. En route, the pair couldn't resist putting their aircraft to the test, indulging in low flying and 'beating up' airfields, which Eric described as 'the basis for a valid assessment of any finer points of this sedate, good-natured dowager of an aeroplane'. This impromptu handling earned both a reprimand and a three month-reduction in seniority endorsed in their personal records.[5] If there was

5 Eric's naval records show that he did not always conform. On 1 April 1941, he was in trouble with authority after making an unauthorized flight and taking an aeroplane without consent. The Swordfish which he borrowed was flown to Belfast, not for the first time, so he could visit Lynn. He was caught on this occasion because the weather closed in, and he could not make the early-morning return flight across St George's Channel to Scotland and arrive before anyone of importance had noticed.

not such desperation for pilots at this time, then the reprimand might well have been even more severe.

Thankfully, the impatient Eric wouldn't have to wait too much longer to see some real action, as for him, and Fletch, the war was about to begin in earnest.

Eric's first love was the Grumman Martlet naval fighter. He would call the fighter a 'bumble bee'. It had heavy machine guns and flotation gear, which twice saved his life. He learned how to land on flight decks, fight and survive the Battle of the Atlantic in this machine.

8

The Battle of the Atlantic

Winston Churchill believed that the Battle of the Atlantic was the most important engagement of the Second World War. It was crucial for keeping Britain alive and fighting, seeing action from the first to the last day of the conflict. It was the one battle that Britain could not afford to lose.

By 1940, in order to feed its population and sustain its industry, Britain needed 20 million tons of food, 12 million tons of crude oil, 200,000 tons of minerals and about the same amount of raw rubber, annually, all imported by sea. Whether it was new alloys, like tungsten and molybdenum for building aeroengines for the new fighter and bomber aircraft, or wood and pulp for paper, or refined petrol or crude oil, they all came in ships.

Until the capitulation of France, many of these raw materials came from mines in French North Africa, and until the invasion of Norway iron ore was imported from Scandinavia, especially Kiruna in Sweden. Now, after the fall of those countries, goods came to Britain from all over the globe; from Halifax at the end of the great Canadian railway system, the Caribbean oil refineries, as well as the Indian Ocean, South America and southern Africa via the Cape. Conversely, goods also went outbound to resupply British outposts around the Indian Ocean and the Mediterranean, including Malta and North Africa. To achieve all of this was a tall order, especially when under attack.

The Germans quickly realized that to force Britain to surrender they had to break its supply lines. Thus, German submarines and bombers prowled the Atlantic, attempting to blow any British or Allied ships out of the water. The initial losses of destroyers caused real problems for the Admiralty and its 'naval control of shipping' enterprise. Even the breaking of the German 'Shark' codes by Bletchley Park did little, especially in 1941, to alleviate the threat to Britain's lifeline. Not for nothing did Churchill write, 'The only thing that ever really frightened me during the war was the U-boat peril.'[1]

Churchill, and a senior naval officer, Captain Matthew Slattery, deputy director of the Naval Air Department in the Admiralty,[2] understood that aviation at sea was the best way of defeating German submarines and bombers. However, by the summer of 1941, so many of

1 Winston Churchill, *The Second World War: Their Finest Hour.*

2 Twenty years later, in 1962, Eric Brown would make this role his own and innovate aircraft types and flight-deck configurations just as radical as those in the early war years.

the Royal Navy's aircraft carriers had been sunk or badly damaged that a new, radical solution, was needed to give air cover to convoys.

As there wasn't the time to build new carriers from scratch, the Admiralty made use of merchant ships, by removing the superstructure and then adding plating to form a flight deck. It was a practical solution, but not everyone was happy with this. In particular, the Ministry of War Transport objected to precious merchant ships being taken out of service to be refitted. Churchill overruled them. He believed that above all else, aircraft carriers were crucial in winning the Battle of the Atlantic, and we had to produce more by whatever means necessary.

This work was usually carried out in American yards, as British ones were fully engaged in the construction of replacement destroyers and merchant ships. Yet as the merchant ships were being converted to aircraft carriers, Britain faced another problem. Since the First World War it had ignored the development of naval aircraft and was seriously lagging behind the Germans and Italians. That was until 1940, when the small, rugged and almost barrel-shaped Grumman Martlet appeared on the scene.

The Martlet[3] was really a modified version of the US Navy's F4F Wildcat fighter, but it had been designed

3 The name was later changed from Martlet to Wildcat in January 1944 to coincide with US nomenclature.

by Grumman to be a fleet fighter, rugged enough for deck landing in the stormy conditions of the Atlantic. It was Eric's good fortune that 802 Naval Air Squadron was one of the first to have them. The squadron's fleet of Martlets was originally destined for the French Air Force, but when France capitulated in June that year, the order was taken over by the British Purchasing Commission in America.

Soon after, the fleet of Martlets arrived in Britain complete with metric instruments and painted in a camouflage pattern, which Grumman factory workers in New York State thought was what the French Air Force required. It wasn't, but there was no time for a repaint before the fighters were needed, as Britain was on the back foot against the Germans at sea and in the air.

At RNAS Donibristle, Eric heard the Martlet arrive before he could even see it. While he sat in the squadron office, a 'whining scream' filled the room, which became a 'thunderclap' as it got closer. Rushing to the window, Eric caught sight of the new fighter that was going to have such an impact on the war, as well as his career. In his words, it looked like 'an angry bee'. He expressed his admiration for the new Martlets in his book *Duels in the Sky*:

> The Wildcat was a great asset to the Fleet Air Arm, bringing it to [almost] the level of the fighter

opposition. It was also an aircraft specifically designed for modern carrier operations, thereby setting new standards for British designers in the field. The Wildcat was a potent fighter, with splendid manoeuvrability, good performance, heavy firepower and excellent range and endurance. On top of this, it was a superb deck-landing aircraft.

While Eric heralded its improved performance, and suitability for aircraft carriers, he also recognized its superior firepower. Previously, the Fleet's aircraft had been equipped with the lighter Vickers or Browning 0.303 inch guns. In contrast, the Martlet was fitted with 0.5 inch Colt-Browning machine guns.

With the pilots of 802 Squadron earmarked for aircraft carrier operations the following summer, they not only needed to master the new Martlet, but also had to learn how to land it on an aircraft carrier. This would prove to be quite a challenge. The pilots at Donibristle were a real mix, as some, such as twenty-year-old Norris Patterson, had already seen front-line service in the Battle of Britain, while others, like Eric, were new to both fighters, as well as landing on carriers.

To train them as quickly as possible, the highly competent and charismatic leader of the squadron, Acting Lieutenant-Commander John Wintour, drove his pilots hard to get their 'sea legs'. Typically, the training started with landings on a dummy deck, marked out with white paint on the runway at Donibristle. From here, the pilots were then able to find their feet training on the recently commissioned training carriers HMS *Argus* and HMS *Victorious*.

However, some pilots found the new Martlet fighter difficult to master. One young pilot lost his bearings, flew into high ground and was killed. Another ditched into Loch Lomond, although the pilot, Sub-Lieutenant Roderick (Sheepy) Lamb, survived. Indeed, Eric faced his own issues.

During just his second flight in a Martlet, a fire caused the engine to stop. Thankfully, Fife's Burntisland Reservoir[4] was just below, leading to Eric ditching the fighter in the water. On impact, the Grumman-patented automatic flotation bags kept the aircraft afloat, leaving Eric to swim clear, with just a small facial injury to show for it. The bags also kept the Martlet upright until it was salvaged and eventually returned to service.

Eric was soon learning that his small stature was a real bonus in a Martlet. Not only did it mean more space in the cramped cockpit, but if it should jolt and jerk, he wouldn't slam his head against the roof, which in some instances could be fatal. Indeed, during this period Eric came to be called the nickname he would be known by

4 Also known as Stenhouse.

for the rest of his life. All Fleet Air Arm pilots who were 'vertically challenged' were nicknamed 'Winkle' or 'Periwinkle', including the Fleet Air Arm Channel Dash hero Lieutenant-Commander Eugene (Winkle) Esmonde VC.[5] As Eric stood at a slight 5 feet 6 inches, he also earned the moniker. From then on, for most in the Fleet Air Arm, there was only one 'Winkle'. In a letter to his father in Surrey, Norris Patterson wrote that 'Winkle had the devil in him at times.' This was the first written record of his nickname but it was one that was to endure until Eric passed away.

Eric might have been small but he was always desperate to be seen – and known – especially to VIPs. He grabbed his chance when he and Norris Patterson were asked to provide an escort for the Prime Minister, Winston Churchill.

As Patterson and Eric were escorting the VIP transport aircraft, Eric couldn't help himself. Inverting his plane, he proceeded to fly upside down, at which a stunned passenger took some photographs. The fighters all had letters painted on their fuselage, and the resulting series of pictures show R for Robert inverted with K for King and Q for Queenie the usual way up. It was indeed quite a show.

5 Esmonde earned his Victoria Cross when his squadron attacked parts of the German fleet attempting a 'Channel Dash' from Brest to their home bases at Wilhelmshaven and Kiel via the English Channel.

Soon after, Eric would again get a chance to catch the Prime Minister's eye. Churchill had taken a keen interest in the Martlet, but after the first ones had been delivered to the Aeroplane and Armaments Experimental Establishment at Boscombe Down,[6] it was reported that they were under-powered and lacked folding wings, an important feature of a naval aircraft if it is to be stowed in the cramped hangar of an aircraft carrier. The PM demanded action on a folding-wing version of the fighter and tasked the Fifth Sea Lord, Rear-Admiral Lumley Lister, the Admiralty Board member responsible for naval aviation, to go to America to find out what was happening. Before Lister's transatlantic visit, Churchill decided he would take him, and the First Sea Lord, Sir Dudley Pound, up to Scotland to see 'the new Grumman fighter'. Having heard that Churchill would be visiting, Eric again tried to make himself known. Memories of his flight with Ernst Udet in Germany just six years before must have been on his mind as he aimed to take off, flip inverted and fly around the circuit thus configured, before lining up and flipping back to right side up for the landing. It would certainly be difficult to pull off, but Eric thought it was worth the risk.

However, with the PM watching, as Eric went inverted,

6 A&AEE (Aeroplane and Armaments Experimental Establishment) is the government research establishment in Wiltshire for in-service trials and evaluation of aircraft and their weapons.

the Wright Cyclone engine quit with a bang and a flash of flame. The Martlet carried on flying for a few minutes but then crashed into the Firth of Forth. Although the Grumman flotation gear again worked, keeping the fighter above sea level, it was upside down, with Eric strapped into the underwater cockpit. He counted his blessings that the cockpit had remained intact, preventing any water rushing in. Rather than water in the cockpit, he was instead aware of a lot of blood. The impact of the crash had caused Eric to break his nose and bruise his right arm. Taking a few moments to gather his wits, Eric unstrapped himself, opened the cockpit, and swam out, before being picked up by the crash-boat and transferred to hospital.

Typically, Eric used the experience as an opportunity to reflect and improve. 'This was when I realized that to survive,' Eric recalled, 'I had to practise underwater escape from a confined place and to do it in the dark. I would do that by hanging upside down in a garage, at night, with the lights out.'[7]

Despite the fact that Eric had written off a valuable fighter, Churchill singled out the CO in the mess and sent Eric his condolences. In later life, Eric would sometimes claim that Churchill came to his bedside, but there is no record of this happening.

7 Conversation with the author in 2012.

Throughout all of this, Eric continued his relationship with Lynn Macrory. After he had left Northern Ireland, much of their relationship had been confined to letters, but his posting in Scotland now made it far easier for him to travel to Belfast, via the ferry crossings at Larne, or even on some occasions by borrowing an aeroplane. However, on 15 April 1941, Eric feared he had lost her.

During the night, the Luftwaffe delivered a bombing raid on the city that became known as 'Easter Eggs for Belfast'. It caused terrible fires in the city centre and resulted in multiple casualties. Unable to make contact with Lynn, and fearing the worst, Eric made his way to Belfast, where amidst the burning rubble he was relieved to find that she had escaped unhurt, her house untouched, and along with her family was doing all she could for her community, even helping to set up the aid centre in the Fitzroy Presbyterian Church.

The attack was to prove a major own goal for the Germans. Until this point the anti-British Irish Taoiseach, Éamon de Valera, had seemed in danger of siding with the Nazis, potentially offering the Luftwaffe and Kriegsmarine an opportunity to use neutral Ireland as a base to attack the United Kingdom. Yet the attack against Belfast caused such outrage throughout Ireland that all relations with the Nazis were immediately cut off.

As an orphan Eric had longed for a family of his

own, and now that he had found the woman he hoped to one day marry, the thought of losing her was almost too much to bear. Not wanting to waste any more time, on 17 April 1941 Eric proposed. As well as presenting Lynn with a ring, he also gave her a spaniel named Gulliver. It seems that this helped to seal the deal, as Lynn said yes.

But there was no time to celebrate, as Eric had to return swiftly to training in Arbroath, where 802 Squadron were readying themselves for their posting to an aircraft carrier.

After practising landing into a set of arrester wires across the runway at Arbroath, the squadron were assembled by the CO, John Wintour. He hushed the room and announced that they would soon embark for their carrier, HMS *Audacity*. Eric and company were bemused. They had never heard of it. 'She's called an auxiliary carrier,' Wintour explained, 'and she's different to anything that has sailed before.' This was saying something.

Due to the immediate shortage of aircraft carriers, HMS *Audacity* was actually the captured German banana-boat *Hannover*.[8] The German boat had been converted into an aircraft carrier by cutting down a couple of decks and constructing a steel flight deck in their place.

8 The ship had several names during its short life including *Sinbad*, *Empire Audacity* and, briefly, HMS *Empire Audacity*.

Wintour said that HMS *Audacity* was to be tasked with protecting Gibraltar convoys from attacks by the Germans, particularly from aircraft. After France had capitulated in June 1940, there were a host of good bases available to the Germans from Brittany to the Spanish border, which allowed their maritime bombers to be well within range.

Wintour went on to explain that while *Audacity* would carry just six Martlets, she was also only 420 feet long and 60 feet in beam. This caused some concerned chatter. Other carriers had flight decks that were twice as long. Indeed, Eric and his fellow pilots had trained to land on the longer flight decks. This had proved hard enough, but landing on a much shorter deck was a challenge for the most experienced of pilots, let alone those who were newly qualified.

Wintour sensed the apprehension in the room and approached the blackboard. He explained that to bring the returning aircraft safely on to the flight deck, there would be a Deck Landing Control Officer.[9] This was an experienced pilot who gave visual signals to the landing pilots. The job was especially important before touchdown, when the officer gave the 'cut' or the 'chop' signal to the pilot, who would then retard the throttle, allowing

9 The so-called 'batsman' because he used two yellow bats, rather like table-tennis bats in shape but larger.

the Martlet to float on to the flight deck and catch one of the arrester wires strung across the deck, taking a hook on the Martlet's tail. HMS *Audacity* and 802 Squadron were lucky to have Lieutenant John Parry in the role. He was experienced and fearless – standing in the open, on a small platform to the port side of the flight deck, in all weathers in which the Martlets flew.

However, Wintour explained that there was another challenge. Due to the shorter flight deck, there would be just two arrester wires to catch on landing. Most other carriers had as many as six. He attempted to calm the room by stating that there would also be a final wire for a fighter to catch, called the 'Christ's-Sake-Wire'. The name didn't make anyone feel any better. Wintour said that if a pilot was to somehow miss the first two then this final wire was an added insurance. Although he added that if they were to use the Christ's-Sake-Wire it would probably break the aircraft in two as well as break the pilot's neck. At this the concerned murmurs somewhat increased.

And still the bad news kept coming. Due to the lack of space, there was no hangar in which to keep the Martlets. They would have to be stored out in the open, on deck, where the mechanics would have to work on them throughout the night to ensure they were ready to fly the following day. These brave young mechanics worked in conditions which today would raise the hairs of any health and safety inspector. While exposed to the elements, and just a few feet from the flight deck side, and the ocean below, they had to fix the Martlets using only a small hooded blue torch for light. Anything brighter ran the risk of being spotted by German aircraft and submarines. The mechanics were often unsung heroes of the war, but their pilots certainly knew their worth.

Thankfully, as Eric and his fellow pilots grew increasingly disturbed, Wintour explained that there were some benefits to *Audacity*. As she was a converted merchantman, this meant cabins and baths, rather than mess deck bunks and conventional naval 'heads' with a small shower. It also boasted a fully functioning galley, which meant that both the merchant navy crewmen and those from the Royal Navy were well fed. Indeed, the ship's company included trained chefs, not just plain cooks as in other navies, who delivered as many as seven meals a day for the various watches required in a warship.

There was, however, something Wintour decided to keep to himself. Merchant aircraft carriers like HMS *Audacity* lacked naval-standard damage-control provisions. This made them more vulnerable to rapid-spreading fires and water ingress in the event of being torpedoed. This was clearly on a need-to-know basis, and the pilots already had enough to contemplate.

As Wintour finished his briefing, he looked up to his squadron and asked with a smile, 'Any questions?'

'We sat speechless and aghast,' Eric remembered.

The night before Eric was due to practise his first landing on *Audacity*, and then set sail for war, he was scared. So were his fellow pilots, who all sat quietly in the mess at Campbeltown on the Clyde, barely touching their drinks, before retiring to bed. In the darkness, Eric tried to visualize his approach, and what he would do if he, and the aircraft, were to miss the wires and go over the side. He was somewhat comforted that he knew the Martlet would float, having crashed it into the water before, but that had been in a calm reservoir, not in a raging ocean.

Apprehensive, Eric woke early the following morning, a sick feeling gnawing in his stomach. While awaiting his turn to attempt to land on *Audacity*, he listened to the CO giving instructions, before the call came: 'Winkle, you're up!' Taking off in his Martlet, after fifteen minutes in the air, Eric caught sight of something in the ocean below. It was *Audacity*, looking 'terrifyingly small', bobbing in the water like a bath toy, without even so much as an 'island' on board to act as a lighthouse. Moving any fear to one side, Eric focused on the job. Feeling a sense of calm wash over him, he lowered the Martlet until he was 400 feet dead astern. Proceeding to lower the wheels and hook, he then saw the batsman wave his bats into a 'cut'. It was time to land. At 70 knots, and throttling back, Eric felt a bump as he hit the deck, and then a tug as he caught the first wire before trundling to a stop. He was ecstatic. It was as smooth a landing as he could have hoped for. With his confidence high, over the next few hours Eric practised six more take-offs and approaches, with no incidents.

With landings complete, *Audacity* and her crew set off, protecting a convoy taking ammunition to Gibraltar. Yet while Eric was confident in his own ability to take off and land on *Audacity*, it was still one of the most vulnerable aircraft carriers in the Fleet, especially when faced with the might of the German U-boats and bombers.

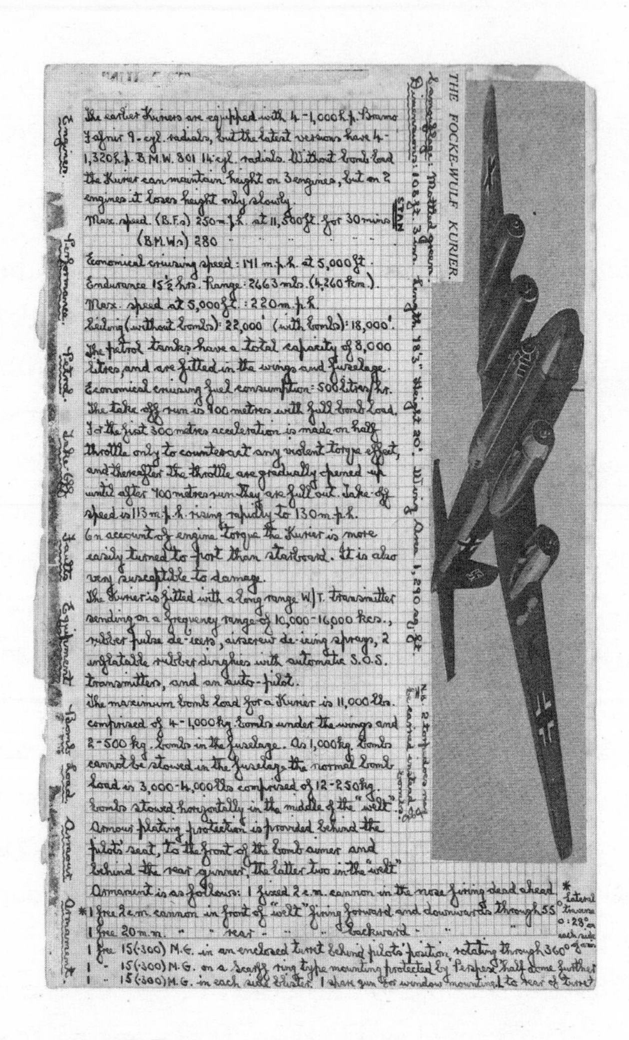

THE FOCKE-WULF KURIER.

Camouflage: Mottled green.
Dimensions: 108 ft. 3 ins. Length 18'3". Height 20'. Wing Area 1,290 sq. ft.

Engines. The earlier Kuriers are equipped with 4 - 1,000 h.p. Bramo Fafnir 9-cyl. radials, but the latest versions have 4 - 1,320 h.p. B.M.W. 801 14-cyl. radials. Without bomb load the Kurier can maintain height on 3 engines, but on 2 engines it loses height only slowly.

Performance. Max. speed. (B.F.s) 250 m.p.h. at 11,500 ft. for 30 mins.
(B.M.W.s) 280 " " "
Economical cruising speed: 141 m.p.h. at 5,000 ft.
Endurance 15½ hrs. Range: 2663 mls. (4,260 km.).
Max. speed at 5,000 ft.: 220 m.p.h.
Ceiling (without bombs): 22,000' (with bombs): 18,000'.

Petrol. The petrol tanks have a total capacity of 8,000 litres, and are fitted in the wings and fuselage.
Economical cruising fuel consumption: 500 litres/hr.

Take-Off. The take off run is 900 metres with full bomb load. For the first 300 metres acceleration is made on half throttle only to counteract any violent torque effect, and thereafter the throttle are gradually opened up until after 900 metres run they are full out. Take-off speed is 113 m.p.h. rising rapidly to 130 m.p.h.

Faults. On account of engine torque the Kurier is more easily turned to port than starboard. It is also very susceptible to damage.

Equipment. The Kurier is fitted with a long range W/T. transmitter sending on a frequency range of 10,000-16,000 kcs., rubber pulse de-icers, airscrew de-icing sprays, 2 inflatable rubber dinghies with automatic S.O.S. transmitters, and an auto-pilot.

Bomb Load. The maximum bomb load for a Kurier is 11,000 lbs. comprised of 4 - 1,000 kg. bombs under the wings and 2 - 500 kg. bombs in the fuselage. As 1,000 kg. bombs cannot be stowed in the fuselage the normal bomb load is 3,000-4,000 lbs comprised of 12 - 250 kg. bombs stowed horizontally in the middle of the "welt".

N.B. 2 torpedoes may be carried instead of bombs.

Armour. Armour plating protection is provided behind the pilots' seat, to the front of the bomb aimer and behind the rear gunner, the latter two in the "welt"

Armament. Armament is as follows: 1 fixed 20 m.m. cannon in the nose firing dead ahead
*1 free 20 m.m. cannon in front of "welt" firing forward and downwards through 55°
1 free 20 m.m. " " rear " " " backward " "
1 free 15 (·300) M.G. in an enclosed turret behind pilots' position, rotating through 360°
1 " 15 (·300) M.G. on a Scarff ring type mounting protected by Perspex half dome further
1 " 15 (·300) M.G. in each side blister. 1 [illegible] gun for window mounting to rear of turret

* lateral traverse: 28° on each side of [illegible]

Eric's first duty aboard HMS *Audacity* was to create the 'threat profile' of the Condor maritime bomber, seen here for the first time in his own hand.

9

First Action

By late 1941, the Germans had almost perfected the disruption of British convoys thanks to the twin deployments of bombers and submarines. The key to this was a heavily armed four-engined maritime bomber called the Focke-Wulf Fw 200C Condor. Eric had seen it in Germany on several occasions in its original incarnation as one of the first long-range, land-based passenger airliners. Since then, it had been converted into a bomber which Eric described as 'very, very potent . . . heavily, very heavily armed'. It boasted a considerable bomb load of over 5 tonnes, an effective machine-gun self-defence and ship-strafing capability from forward-firing cannon. There is no better phrase to describe the Condor's awesome firepower than 'winged porcupine'. From any direction, it was a threat.

Although it could bomb convoys out of the water by itself, it also acted as a lookout for U-boats. On spotting a convoy in the ocean below, the bomber would lurk within sight, but out of anti-aircraft gun range, before signalling back course, speed and details of the escorts to Befehlshaber der U-Boote headquarters in Hamburg, from where the worldwide deployment of German submarines was coordinated. At this stage, the Condors had no radio capable of talking directly to the U-boats, so this information would be passed from Hamburg to any U-boats in the vicinity, alerting the submarine commanders as to where their prey could be found. With this information, a 'wolfpack' interception line could then be created.

With radar not yet operational, the British convoys had to rely on almost continuous air activity by the Martlet fighters to detect and engage the Condors of Kampfgeschwader 40, a squadron based in occupied France. The aim was to keep them from either bombing the merchant ships or acting as the *Fühlungshalter* (tactical coordinator), radioing the convoy's position to base. Ensuring a convoy's safe passage was absolutely crucial to the war effort, so this was a vital role.

Whenever possible, the Martlet fighters were also tasked with engaging any visible U-boats. The early U-boats were better suited to attacking on the surface at night, where the speed of their diesels and better situational awareness for the skipper could prove decisive when attacking convoys. This at least gave the British pilots a chance of seeing them. However, the Martlet's heavy-calibre Colt-Browning machine guns were ineffective against U-boats, with most of the shells bouncing

The Salvation Army's Mothers' Hospital in Hackney is the birthplace of our greatest pilot.

Eric had a great affection for animals, especially dogs.

Robert and Euphemia Melrose Brown lavished attention on their only child.

Robert and Eric on holiday in Germany together in 1937, probably in Ulm. Eric was proud of his scholarship to the Royal High School in Edinburgh.

First World War fighter ace Ernst Udet was an early hero of Eric's. He claimed Udet took him flying.

Nazi aviatrix Hannah Reitsch featured prominently in Eric's thoughts before and after the Second World War.

Eric's initial flying training was centred on the DH Gipsy Moth biplane, in which he first flew solo on 11 September 1938.

Eric's aviator's certificate, showing an earnest young man of 18 – already a promising flyer.

Contrary to the popular belief, Eric carried out his pilot training in England under the Civil Air Guard programme. There was no University Air Unit at Edinburgh before the war.

Eric was on a driving holiday in Germany when war was declared in 1939. After being arrested by the Sicherheitsdienst, the Nazi security police, he was deported to Switzerland but allowed to keep his beloved MG Magna L2.

Eric found that the Royal Air Force had no room for him in late 1939, but that the Royal Navy was rapidly building the Fleet Air Arm. When he applied, he found his previous training was to no immediate avail, and he would still have to practise 'square-bashing', marching drills on the parade ground.

Eric's first passport, including the German visa issued in July 1939 and the entry stamp for 1 August.

Eric made friends easily in the Royal Navy. The training brought together people from all walks of life and he found he fitted in.

Eric made his combat debut against the Luftwaffe in the autumn of 1940 flying the obsolete Gloster Sea Gladiator biplane.

802 Naval Air Squadron was the second unit to be equipped with the American Martlet fighter, originally destined for France until that country's collapse.

Ordered to escort Winston Churchill's aeroplane to Scotland, so the Prime Minister could see the Martlet fighter, Eric couldn't resist the opportunity to stand out, flying upside down. This picture was taken by one of the Prime Minister's party.

Captured in the Caribbean and modified in Britain, the banana-boat *Hannover* was to become the prototype for a whole new class of warship – the escort carrier.

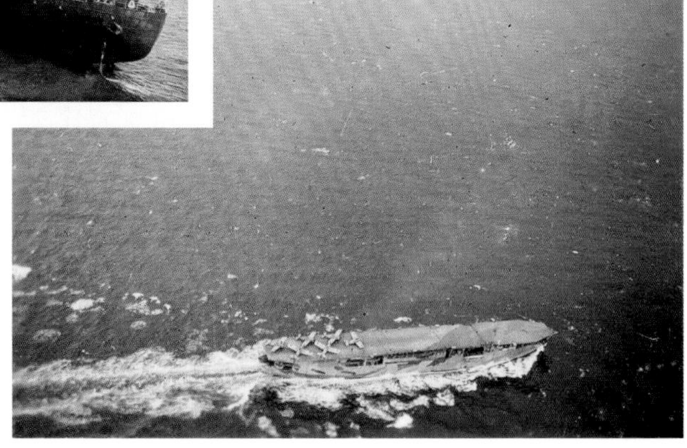

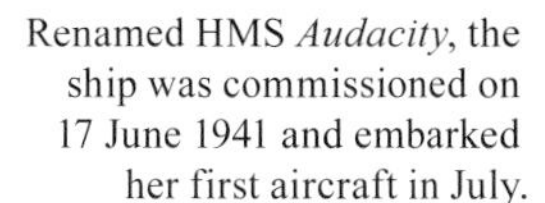

Renamed HMS *Audacity*, the ship was commissioned on 17 June 1941 and embarked her first aircraft in July.

HMS *Audacity*'s short life was a useful one, escorting convoys to and from Gibraltar at a crucial time of the war.

Eric's 'stick buddy' in 802 Naval Air Squadron was Graham Fletcher, seen here in the cockpit. Fletch was sadly lost when HMS *Audacity* was sunk in December 1941. He would have been Eric's best man. The other officer, on the cowling is probably Norris (Pat) Patterson. The capless officer is John Sleigh.

An 802 Naval Air Squadron Martlet prepares to launch from HMS *Audacity*'s tiny flight deck.

Publicity photograph of Eric in the cockpit of a Martlet aboard HMS *Audacity*, wearing the newly installed Sutton shoulder harness. These straps saved his life when he was wounded in an attack on a Condor.

Eric joined the Battle of the Atlantic in September 1941 when 802 Naval Air Squadron first embarked on HMS *Audacity*. Appointed the air warfare officer, Eric set about finding weak points to counter the German bombers plaguing convoys. It helped that he had seen the Condor in Germany in 1937, but this didn't stop him nearly losing his life to its defensive armament.

Michael Turner's amazing artwork of HMS *Audacity* sinking as described to the artist by Eric. By this time, Eric was already in the water swimming clear.

How Eric would have looked stepping off HMS *Deptford*'s prow at Liverpool in his service dress whilst many other survivors of HMS *Audacity* were in rough battledress, looking like enemy submariners.

off the steel hull. Despite this, the mere presence of an aircraft could cause a U-boat to crash-dive, and thus its speed would slow by about 10 knots, often making it impossible to catch the convoy.

In his first few days at sea, Eric was full of adrenaline and ready for the fight. Yet nothing happened. His days were spent endlessly patrolling the empty grey skies, feeling lonely and bored. Radio was not allowed, so he couldn't speak to his fellow pilots, while the lack of action made him think this might not be as exciting as he had hoped. However, when a call went out that a U-boat had been spotted on the surface, Eric eagerly charged towards it. Finding it unprepared, as it was charging its batteries, Eric and his leader fired 400 rounds until the submarine crash-dived. The damage to the submarine's structure was minimal, but the German crew caught on deck when it dived all perished. For Eric's first engagement with the enemy at sea, it was a gentle introduction, and at least allowed him the thrill of opening fire. The following day, however, things would get a little more intense.

On the night of 20/21 September, U-boats sank three ships in the convoy, with survivors put aboard the Red Cross rescue ship the *Walmer Castle*, as well as *Thames*, a rescue tug. The following morning, with the ships still looking for survivors, a Focke-Wulf (Condor) bombed the *Castle*, despite its clear Red Cross markers.

The ship erupted in flames, with many of the survivors being burned alive. This was an outrage, even during war. It was well established for both sides that Red Cross ships were off limits.

In response, two Martlets were immediately scrambled from HMS *Audacity*, with orders to intercept the Focke-Wulf and take vengeance. Fletch and Norris Patterson were the first to set sight on her, just as the German bomber was preparing to drop its load on the *Thames*. With no time to waste, they opened fire, shooting thirty-five rounds from each gun, blasting the German bomber's tail unit off and sending it hurtling into the ocean below. Nobody aboard HMS *Audacity* mourned the German pilots' passing. It was unacceptable to bomb a hospital ship, and those responsible were considered to be no better than pirates.

Back aboard HMS *Audacity*, so delighted were the mechanics that Patterson and Fletch were plied with a heady mixture of navy rum and water, known as grog. Eric and the other pilots also celebrated with a rare pink gin, the naval officer's drink of choice at sea. It was unusual for aircrew to drink at sea, but this was an important occasion, which needed to be marked in an appropriate fashion. In a letter home, Patterson recalled that the squadron pilots 'cut my black naval tie, about an inch below the knot. It was some sort of celebration, but I wish they hadn't.' It was the sign of a combat victory, and a practice which continued on aircraft carriers for years.

However, at noon, the Germans were back. The *Thames* had previously enjoyed a lucky escape, but this time it was bombed by a Junkers 88, which escaped before it could be intercepted. German attacks followed throughout the night, with three more ships torpedoed and HMS *Audacity* taking on eighty-six wounded survivors. The ship's doctors were so overworked that Eric was asked to lend a hand as an amateur anaesthetist. Looking on at some of the wounded men, blood everywhere, screaming in pain, Eric finally realized what war was really all about. It was a baptism of fire.

The rest of the trip was thankfully nowhere near as eventful, especially the latter part, which saw Gibraltar-based patrol aircraft take over convoy escort duties. Bored of waiting for more action, Eric and Wintour decided to 'take a look' at the southern part of neutral Portugal and non-belligerent Spain. It seems to have been a simple overflight at a few thousand feet, where there was little risk of being shot down by ground fire and no real risk of fighter interception. It was not, though, the most diplomatic or strategically sensible approach, when the British government was trying by every means to keep Spain from joining the Axis nations. Despite this, Eric remarked that it was 'truly thrilling', and no doubt it broke the monotony of waiting around or flying fruitless patrols over water.

On 27 September 1941, HMS *Audacity* tied up alongside in Gibraltar, while the Martlets landed at the local Royal Air Force station at North Front for a thorough inspection. The fighters had stood up to the extreme conditions exceptionally well, and only minor repairs were needed. Most importantly for the pilots, shore leave was granted and taken. This gave Eric and his messmates the opportunity of letting off some steam with a few drinks, as well as sending letters home to loved ones. Meanwhile, merchant ships from ports in the eastern Mediterranean and from Freetown in Sierra Leone gathered in Gibraltar, ready for HMS *Audacity*'s return trip to Britain.

By 2 October, when the convoy departed for Scottish waters, there were seventy ships to be shepherded across the Bay of Biscay, which was being lashed by some of the worst equinox storms ever recorded. As the weather was so atrocious, the Martlets were unable to launch, so a Coastal Command Catalina flying boat and a Hudson patrol bomber, both from Gibraltar, provided a covering force for the first day or so. Even when the Atlantic swell abated, the aircraft carrier's stern rose and fell 65 feet, as the ship rolled as much as 16 degrees off its datum, as measured by the ship's sextant.

Nevertheless, as HMS *Audacity* moved north into the Bay of Biscay, regular patrols were required to take off.

With the waves rising high and crashing over the deck, the take-off start was judged, and signalled, by the highly experienced Captain, Commander (P) Douglas MacKendrick.[1] With his naval cap, rather than traditional flags, MacKendrick signalled the pilot to start his take-off run just as the ship started to pitch up, thus sending the Martlet into the air at exactly the right moment to climb away above the waves. With every take-off there was a real danger of an aircraft smashing into the wall of water and being dragged into the ocean below. It was certainly not for the faint-hearted.

With sightings of U-boats and Condors en route to the Clyde, the Martlets were frequently scrambled. As described in the Prologue to this book, one such 'scramble' saw Eric almost lose his life, but also make his reputation.

Following a Condor being sighted on the edge of the convoy, Red Section was ordered to take off, with Sub-Lieutenant Sheepy Lamb in the lead and Eric on his starboard wing, tucked in close. The two Martlets closed at best speed of 280 knots, or 5 miles a minute, setting the gun button to fire, keyed up for action. Yet as they engaged with the Condor, Eric's Martlet was hit, with plexiglass shredding his mouth, and he was rendered

1 The 'P' stands for 'Pilot' and shows that he was familiar with ship handling and flying.

unconscious. With Eric out of action, his Martlet was not only an easy target, but was also in danger of falling into the ocean.

As Lamb watched on, helpless, Eric's eyes suddenly fluttered open. He turned to Lamb and immediately recognized the seriousness of the situation. He had to get back to *Audacity*, and fast. However, with blood in his eyes, he couldn't see where he was going, let alone land on the violently moving deck of the carrier.

While Lamb led the way back to *Audacity*, and tried to navigate, Eric took his left hand off the throttle and moved it to the stick, so the right hand could spin the undercarriage wheel, bringing the heavy and cumbersome undercarriage down by hand. This was vital. If not set up properly, once the locking mechanism was disengaged, it could run free and damage the undercarriage. It could also cause injury to the pilot, who would need his right hand to land. It was never an easy procedure at the best of times, but when wounded, and partially sighted, it took courage, strength and sheer determination; luckily 21-year-old Eric had plenty.

Flying down the port side of the carrier, turning short in a rounded approach to line up on the jerking flight deck, Eric desperately looked for the yellow bats of the deck landing control officer. Somehow, he caught a flash of them through the driving rain and made his approach.

Battling against losing consciousness, Eric managed to keep the Martlet on the right approach track, hitting the deck at speed, charging past the first two wires, before taking the Christ's-Sake Wire, which was extra taut and brought the Martlet up hard. Thankfully, the Martlet stayed intact, but Eric remembered too late that his shoulder straps were loose, as he liked them that way for air combat, so he could twist his body around to keep the enemy in sight. On catching the wire, he jolted forward, bending almost double, smashing into the gunsight, collecting more glass and metal fragments in his face and knocking himself unconscious.

He awoke soon after in the ship's sick bay, his face bruised and bloodied, but he was alive. In the most desperate circumstances it had been a quite miraculous feat of flying, and it certainly did not go unnoticed. In his official report, Wintour described the landing as an 'outstanding achievement'. This comment was endorsed by the ship's Captain, Commander MacKendrick. Both praised Eric for his skill and natural ability to land an aircraft in the worst set of conditions possible. It was clear that Eric was a natural.

Rather than celebrate his achievement, Eric felt a sense of shame. He felt he was the air warfare instructor on attacking the Condor and had been the first pilot to almost fall victim to the bomber's firepower. As well his pride being dented, he was also off flying and kept in the ship's sickbay until the convoy reached the Clyde, with no ships lost.

He kept a souvenir of the Condor's efforts for the rest of his life – a piece of the armoured glass remained lodged between his jaw and palate, as it was too difficult to remove. This sometimes made chewing difficult and occasionally caused Eric real discomfort. More than once he was known to have repaired cracked partial dentures with Superglue, especially when about to give a lecture and the wartime fix broke.

However, on Eric's arrival in Scotland, he found that the Admiralty had received word of his feats, and they were now keeping a special watch on his talents. But before Eric had any thoughts of greatness, he almost lost his life.

HMS *Audacity* had a short operational life. She was the number one target for Germany's U-boat wolfpacks because her aircraft were so effective against bomber and submarine alike. On 21 December 1941 the escort carrier was sunk by torpedoes from U-751, probably on her first war patrol. Eric was lucky to survive, partly through the good sense of taking his aircrew life jacket when he jumped overboard and partly because of the warm clothing he was wearing.

10

German Bombardment

When British Convoy OG 76 sailed from the Clyde on the night of 28/29 October, its journey was immediately fraught. Not only were the Germans tracking it from the minute it rounded the Irish mainland, but on board HMS *Audacity* there were real problems.

For this trip, HMS *Audacity* had been loaded with two extra Martlet fighters and two more pilots, bringing the totals to eight and ten respectively. However, just days into the journey, such was the ferocity of the weather that Norris Patterson's Martlet was lost as it was tossed overboard, with Patterson still inside. Luckily, he was saved by an escorting destroyer. Yet with mechanical issues affecting the remaining Martlets, *Audacity* was soon down to just four serviceable fighters. Things then went from bad to worse.

Several miles from the convoy, a Vichy French Latécoère flying boat was spotted by a merchant ship. With HMS *Audacity* short of Martlets, there was no time to hang about. They had to strike first, and hard, with Eric sent to engage with what was thought to be the enemy.

Events are sketchy following this. Eric claimed responsibility for blowing it out of the air, but it seems that what they thought was a German aircraft was in fact full of French civilian refugees en route to neutral Spain or Portugal.

All of Eric's published works in later life are silent on the event. The only reason we now know of it is because his logbook indicates that there may have been some internal inquiry after it was found that fourteen passengers and crew had died in the incident. This must have been a shattering blow to Eric. He might have reasoned there was a war to be fought, and convoys to protect, and any aircraft coming close had to be intercepted. Nevertheless, it would have hurt knowing that innocents had died. Such is the cruelty of war.

The return trip to Scotland was to prove uneventful, although Eric did manage a quick visit to see Lynn in Belfast. However, the second convoy departure from the Clyde was to become a relentless running battle with submarines and bombers across the Bay of Biscay.

On 9 November, a pair of replacement Martlet fighters, fully fuelled, armed and serviced, were ready for action on the heaving deck. The squadron CO, John Wintour, with his wingman, Sub-Lieutenant David 'Hutch' Hutchinson, were also set and strapped into their cockpits, ready to intercept a Condor on the margins of the convoy, about 25 nautical miles away.

Hutch was an outstanding pilot. He had been posted to No. 74 (Tiger) Squadron in Fighter Command, flying Spitfires in the Battle of Britain shortly after being awarded his wings in June 1940. Later, he flew Sea Gladiators with 804 Naval Air Squadron from Hatston on Orkney and was then one of the first Martlet pilots. He was therefore already an accomplished fighter pilot, even before embarking on HMS *Audacity*.

Following take-off, the Condor was soon spotted through the rapidly deteriorating weather, which meant visibility was down to just a few miles. As Wintour and Hutch sped towards the German bomber, unusually, the pilots' radio was played through *Audacity*'s main broadcast, so that the crew could hear the combat. Over the broadcast, Wintour was heard to say, 'I'm coming in astern now.' Then there was a cry, and static. A cannon shot had ripped through the side of Wintour's Martlet and killed him instantly.

Moments later, Hutch was able to find a successful attacking position and sent the bomber crashing into the sea about 20 nautical miles from the carrier – close enough for those on deck to witness and no doubt to cheer. Another pirate sent to the bottom. But the ship's company, especially the mechanics and aircraft handlers, were stunned by the loss of the CO, who had established a reputation as a good commander and leader of men.

As it looked like a serious battle was developing, a second pair of fighters was brought to readiness. These were the last remaining Martlets, and one of those had a bent airscrew. Eric elected to fly it as he had other things on his mind than the risks. As he put it, 'We were all in an angry mood.'

With Sheepy Lamb leading, the pair launched to the sound of an air-raid warning. Once airborne, Eric tucked in behind his leader's starboard wing, scanning the sky ahead and to the right. Within minutes, they sighted another aircraft skirting through the clouds. However, what at first appeared to have been a Condor turned out to be a large Boeing 314 flying boat, the *Dixie Clipper*. It was unmistakably a 'neutral' civilian airliner, as the markings of Pan American Airways, with the serial NC 18605, were clearly visible. Despite being en route to engage the enemy, Eric could not resist in some aerobatics to entertain the Clipper's passengers, even if his Martlet was in a state of disrepair. A witness to this event described how things almost went badly wrong:

> Feeling the need to relax, [one fighter pilot] crept up on it from astern, dived under it, and pulling up in front of the pilot's cockpit performed a series of slow rolls for the edification of the passengers. That this was reprehensible must be admitted, and unfortunately during the exhibition one of the Martlet's

> machine guns went off. The effect on the Clipper was electric. The pilot fired a red Very light to warn the Martlets off, and an old lady was seen [leaning out] at one of the cabin windows violently waving a pair of white bloomers, which were torn out of her hand by the slipstream and wound themselves round the wing of a Martlet. The pilot returned to the carrier with the [squadron's] first trophy.[1]

With the momentary fun over, Eric resumed the search, almost immediately calling out, 'Tally ho!' Up ahead was a pair of Condor bombers in formation. 'You take the low one, I'll take this one,' Sheepy cried. Making a dive, Eric made three attacks on the leader, first on the port side, then the starboard beam, and finally back to the port side. The bomber tried to escape into cloud, but Eric hung on to its tail like a terrier, keen to avenge Wintour. The opportunity for revenge came very quickly, as the hunted and the hunter came out of the cloud. On a hastily scribbled note, Eric described the resulting conflict in the following words:

> EA suddenly came out of cloud 500 yards dead ahead. Delivered head-on attack, opened fire at 400 yards

1 This remarkable account comes from *The Fleet Air Arm: The Admiralty Account of Naval Air Operations*, published in 1943 by Her Majesty's Stationery Office. It was probably written by Kenneth Poolman, the Admiralty historian and author.

> and broke off at 70 yards. Perspex around two German pilots shattered and nose of EA came up. Took violent avoiding action and saw explosion in nose of EA, possibly cannon fire. EA stalled and went crashing into sea. As second and third attacks were made EA altered course to change attack into full deflection attack. Port wing came off. Escape hatch on top of fuselage in line with trailing edge of wing was open and two crew climbed out. FW filled up and sank. Two survivors lay on floating wing.

The ship's combat log says that the Condor spun in from 1,000 feet, witnessed by those on the aircraft carrier about 9 nautical miles away. Rather chillingly, and reflecting the mode of the time, the two survivors were left in the water to their fate.

Eric and Lamb did not return to the aircraft carrier, as they were close enough to land at Gibraltar. There was no more action, which was just as well because there was only one serviceable Martlet aboard HMS *Audacity* by this stage.

However, the *Dixie Clipper* incident turned out to be a little more of a diplomatic incident than Eric expected. A passenger on board took a photograph of the rapidly approaching fighters and claimed to have seen muzzle flashes from one of the Martlet's 0.50 inch machine guns. The picture was passed to the US Navy, hence to the US Embassy in London, and on to the desk of the Fifth Sea Lord, Rear Admiral Sir Lumley Lister, responsible for the Fleet Air Arm. He was not amused. America was neutral at the time and had only recently seemed likely to join the Allied cause. In a signal, Lister reminded Eric that this was not the first time he had been reprimanded for 'high spirits' and he should 'keep them to engage the enemy rather than for scaring neutrals'. Eric's record was duly marked.

To replace the much-loved and admired Wintour, the Admiralty dispatched Lieutenant Donald Gibson to take his place. Gibson was already a veteran of the early Fleet Air Arm engagements in the Norwegian campaign, as well as operations against the Vichy French in North Africa and the Battle of Cape Matapan in March 1941. Yet he very nearly didn't make it to Gibraltar.

While ensconced in a Sunderland flying boat, en route to Gibraltar to take command, his transport was attacked by German Junkers Ju 88 long-range fighters. As the Sunderland banked and weaved from side to side, to avoid the potent German aircraft, shells passed through the cabin in which Gibson was sitting, scratching him but causing no wounds. Gibson was always considered one of the luckiest naval aviators of his generation, surviving the war in which so many others from his flying course had died. He rose to become a Vice-Admiral and would keep appearing in Eric's service life for the next thirty years.

With Christmas approaching, the ship's company took the opportunity of a month's shore leave in Gibraltar to stock up on presents. The trade was brisk, and Eric was delighted to find a black negligée for his fiancée Lynn. Others bought fruit unobtainable in Britain due to rationing, much of which had been brought over from Tangier in neutral Morocco. Other offerings came across the land border from equally neutral Spain. There were mixed feelings about dealing with Spain, as many suspected that Fascist sympathizers were reporting naval movements in and out of Gibraltar to the German Embassy in Madrid for onward transmission to Berlin. Eric would soon find this to be true.

Presents for the ship's company and air group there might have been, but there were no replacement airframes. This was a real issue. When HMS *Audacity* sailed on the night of 14 December, the HG 76 convoy[2] was large, with thirty-two merchantmen home bound, and sailing in five columns. The main escort was the Second Support Group of five destroyers and seven corvettes, itself a highly practised and potent force, but especially when also supported by an aircraft carrier. However,

2 HG means homeward from Gibraltar; OG is outward to Gibraltar.

Audacity had only four immediately flyable Martlet fighters for the nine pilots on board.

This lack of fighters troubled the convoy commodore, Vice-Admiral Raymond Fitzmaurice. He was, however, glad to have some initial assistance from the radar-equipped Fairey Swordfish aircraft of 812 Naval Air Squadron, flying from RAF North Front at Gibraltar. They patrolled the convoy's flanks for most of the first night, and into the next day, as they steamed south and then west to avoid known submarine locations.

However, after turning on to a northerly course, well off the Portuguese coast and Cape St Vincent, the convoy was picked up by two Condor bombers, flying at extreme range. They scored their first success in the predawn light, guiding a U-boat to a straggling oil tanker in the convoy and sinking it. But when daylight came on 17 December, it was clear that the battle had only just begun.

In the late morning, one of the escorts directed a patrolling Martlet on to a surfaced U-boat. Commanded by Fregattenkapitän Arend Baumann, this was the new Type IXC long-range U-131 on its first war patrol. A state-of-the-art submarine, it boasted a deck-mounted 105 mm gun, while far more dangerous to the attacking Martlet was its quadruple 20 mm anti-aircraft gun-mounting.

Despite this, Fletch damaged the submarine so badly

that it could not dive, and it was scuttled later, at 13:30, with the crew surviving. Fletch was not so lucky. The Quad-20 machine guns brought down his fighter when shells smashed into the cockpit. It is thought they severed the control column and his arm, probably killing him outright. The Martlet dived straight into the sea and sank, but not before the body of the young flyer had floated free. Fletch's body was recovered, and he was later buried at sea, with full honours, his body shrouded in the Union Flag.

It was a loss Eric found hard to take. He had lost a good friend, as well as his designated best man at his fast-approaching wedding day. He had always struggled to make close friendships – as an orphan, he found it difficult to trust and be open with people, especially in the Royal Navy – but in Fletch he had found a kindred spirit. That Fletch had been chosen as his best man goes to show how much he meant to Eric. However, in war, loss of this kind was somewhat to be expected, and there was never much time to grieve, even at the death of a good friend.

By now, it was clear that the Germans had positioned a reinforced wolfpack of submarines directly on the line which the convoy planned to take. From later interrogations of the survivors of sunken German submarines, it was found that the Spanish coast watchers had indeed reported every ship by name and type as they left Gibraltar.

Over the next few days, the battles intensified. Two more U-boats were sunk on 19 December, but it was clear that the Germans were far from finished. Before first light, U-574 sighted and engaged HMS *Stanley*. Seeing the attack on the destroyer from the bridge of his sloop, HMS *Stork*, Commander Johnny Walker initiated a prearranged and rehearsed plan, code-named 'Buttercup'. This saw the escorts disperse, firing snowflake star shells, which illuminated the U-boat, still running on the surface. As she crash-dived, HMS *Stork* went to 'ramming speed' and hit the submarine just aft of the conning tower, then fired a shallow pattern of ten depth charges.

This action, although successful for the escorts, allowed another member of the Seeräuber wolfpack, U-108, to infiltrate the merchant ship columns. Firing torpedoes, it sank the mixed-cargo ship *Ruckinge*, en route to Oban with foodstuffs and agricultural chemicals. Another submarine, not identified, also manoeuvred into position to fire its torpedoes at HMS *Audacity*. Thankfully, the torpedo wake was seen and countered with an emergency helm-over. So far, the ship's luck was holding.

On the flight deck, 802 Naval Air Squadron's heroic mechanics worked through the night, fighting against the stormy seas breaking over the sides. Thanks to their efforts, when three Condors were spotted flying in from the direction of France, six Martlets were serviceable, and they launched.

The first engagement resulted in Eric shooting one down with a head-on attack. But it was not a one-sided battle. Most of the Martlets received hits, and Eric's machine 'had the cockpit hood completely shot to bits and numerous other hits' – he counted thirty-four cannon holes alone. The Martlet was certainly showing that it could take punishment and was growing in Eric's affection.

The wreckage of the first Condor was still floating in the late afternoon when the convoy sailed by, but there was no respite. Another Condor attack was thwarted and a second Condor bomber chased away. When a third appeared in the gathering gloom, it was engaged head-on, first by Sub-Lieutenant Williams and then by Lieutenant Sleigh, who shot it down.

It had been a close combat, and Sleigh landed at last light with, according to the after-action report, a wing tip of one of the Condors hanging from his tail, having hit the German during a close firing pass. Again, the ruggedness of the Martlet was amply demonstrated. It had been a spectacular day.

That night, the officers had a small but, thought Eric, 'terrific' party. Yet the naval ratings working on the maintenance of the Martlets had no time to celebrate. They were once more up all night, out in the open, and working by the light of blue hooded torches. By first light of 21 December, what the Royal Navy called a 'maximum effort' of Martlets was again ready for the fifth day of the continuing battle against bombers and submarines. They did not have to wait long to see action.

After breakfast, Eric and Sheepy Lamb were airborne to the west of the convoy, with Eric taking the port and stern quadrants, while Lamb took the starboard side, searching for U-boats. After two hours, Eric could barely believe his eyes. Two submarines were lying stationary in the Atlantic swell, about 25 nautical miles behind the convoy. The submarines, both probably the long-range Type VIIC, were busy transferring men and supplies across a gangplank, but upon seeing Eric, one opened fire with its powerful Oerlikon anti-aircraft guns. Eric simply climbed and circled for twenty minutes to evade the gunfire until his patience ran out.

Putting the Martlet into a screaming dive, he fired 400 rounds from the Colt-Browning machine guns. The Germans seemed unprepared for the flying assault, and at least three sailors were killed in the fusillade. Whether their task was complete or not, they headed off, on the surface, at about 18 knots to the north-west and into the gloom. HMS *Deptford* and three other escorts were subsequently detached to chase, while a relief Martlet was made ready aboard HMS *Audacity*.

At 11:30, two U-boats were sighted on the surface

astern and to the west of the convoy. Martlets were directed to them, but by the time the fighters arrived, the submarines had submerged. At least the presence of the fighters had hampered the German skippers in their shadowing and search for a position from which a torpedo attack against the convoy would be possible.

During the afternoon, the escorts were almost continually in contact with submarine 'echoes' on sonar. These 'contacts' were attacked with depth charges, which led to the sinking of U-567, killing all forty-seven crewmen, including her Captain, Kapitänleutnant Engelbert Endrass.[3] One down, but five other submarines of the Seeräuber wolfpack were still active.

At last light, about 19:20, Lamb and Eric returned to the carrier. Eric had been the first pilot to launch that day and was the last to land. He had logged around thirty hours of combat time escorting that convoy. It had been a long and momentous day, but it was not over yet.

With darkness providing a cloak around the convoy, merchant ships were still reporting submarines, real or imagined. HMS *Audacity*'s Captain, Commander MacKendrick, decided he would take the carrier out of the convoy, so as to then distract the German submarines by staging a mock battle, with depth charges and star shells.

3 Endrass had been the first watch officer when his previous submarine had sunk the battleship HMS *Royal Oak* in a daring attack at Scapa Flow in October 1939. He was a hero of the German submarine service.

Right on time, at 20:00, the escorts, led by HMS *Deptford*, launched star shells to try and fool the submarines, who liked to close in on a battle in the hope that a wounded merchantman might fall easy prey. It nearly worked, but suddenly one of the merchant ships near the dark shape of the departing HMS *Audacity* accidentally let off a *snowflake* flare, which immediately illuminated the aircraft carrier. To add to the near-daylight conditions, many of the merchant ships, seeing a new mock battle, fired additional snowflakes and star-shells. This provided a perfect aiming point for the submarines, even in the December weather, where sea-level visibility was down to a single nautical mile.

While the sky lit up around the carrier, Cyril Coventry, a Fleet Air Arm Telegraphist/Air Gunner, or TAG, was in *Audacity*'s forward mess playing tombola with two messmates from 807 Naval Air Squadron, Jack Thistlethwaite and Charles McCullagh. As they played, U-751, commanded by Kapitänleutnant Gerhard Bigalk, rose to the surface and, upon seeing *Audacity* illuminated by the flares, let rip a torpedo.

As the men finished playing just their second hand, the ship rocked, and there was a terrific crash at the stern. HMS *Audacity* had been hit at port side, aft, at the engine room, and the carrier immediately started taking in water. Having been built as a merchant ship in Germany, the aircraft carrier lacked naval-style damage control, with watertight compartments in the hull. The water subsequently rushed in, and the ship started to sink.

Incredibly, the men in the forward mess had been in a similar predicament only recently. On 13 November they had been aboard HMS *Ark Royal* when it had sunk and were only on HMS *Audacity* to hitch a lift home. They were not going to be caught on a sinking aircraft carrier again. Coventry and Thistlethwaite therefore grabbed their lifesavers and took the first companion way to the upper deck. In the darkness, Thistlethwaite made out the U-boat which had just opened fire. It 'appeared on the surface only about 200 yards away', he recalled forty years later, 'it was glowing with phosphorescence in the dark and it signalled (by light in English) "abandon your ship" and then submerged again'.

Official and personal accounts differ on whether the German submarine signalled an 'abandon ship' message before continuing its engagement with the stricken ship. Some submarine captains were still operating 'prize rules' of allowing crews to abandon ship before they were sunk. It would have made no difference to the response U-751 received from HMS *Audacity* or its escorts. Their duty was clear: sink the submarine.

A fusillade of light anti-aircraft gunfire from the flight-deck sponsons was aimed at the submarine, but the rounds 'went high as the weapons were designed to fire at aircraft not an object on the surface at close range'. The submarine submerged to escape the gunfire and to load more torpedoes. This process was, however, delayed. In a later German propaganda radio broadcast, her skipper explained that because she was just out of the French submarine base at La Pallice, fresh produce was stored in the torpedo room, and this had to be cleared before reloading could take place. Once the produce had been cleared, two more torpedoes were launched at HMS *Audacity*.

Coventry, Thistlethwaite and McCullagh 'could see the phosphorescence trails coming towards us, so we lay down on deck ready for explosions'. On impact it felt as if the ship had been split in two, as the bow structure collapsed and the stern rose into the air.

Commander MacKendrick ordered abandon ship, and his first action was to throw the ship's confidential books overboard. On hearing the alarm, Eric's first action was equally important. From the wardroom, where he had been drinking coffee and chatting to Lamb, he raced down to his cabin. Just as the lights failed, he took his service cap and aircrew 'Mae West' vest off their pegs and rescued Lynn's Christmas present, then put on his flight jacket and boots. There was, however, no time to rescue his pilot's logbook from the Ops Room. The ship was going down fast, and he needed to reach the deck before he was trapped below.

Racing down the dark corridors, to the sound of the ship's metal cracking and buckling, he finally reached the deck, only to nearly be blown overboard as the aviation fuel tanks exploded.

Rising to his feet, he saw chaos all around. Five of the Martlets broke loose and crashed overboard, wrecking the ship's lifeboats and killing a number of seamen, who all tried to race clear. One Martlet hung precariously to its lashings, but Eric could only watch in horror, hearing the lashings twang as they broke under the strain, and that fighter too crashed into the ocean below. With a horrifying screech piercing the air, the stricken *Audacity* then split in two, with the bottom half of the ship plunging into the depths of the water.

As the remainder of *Audacity* reared vertically into the air, there was a lot of desperate shouting, as crew rushed for the Carley-floats, which acted as life rafts. But there wasn't enough room for everyone. Messmates Coventry and Thistlethwaite were some of the lucky few who found a space, and their life raft was spotted and rescued by the escort HMS *Marigold*. Coventry had survived, but his service career was cut short because as he leaped into the sea from the sinking aircraft carrier, his head struck the water in such a way as to cause concussion. On his return to Britain, he was subsequently passed unfit for further naval service. He was one of the fortunate ones, as he had served, done his duty and would not need to fight again.

Meanwhile, as HMS *Audacity* sank, Eric had no choice but to jump into the freezing ocean. With his leather flying jacket, vest and flying boots, he plunged 20 feet into the water, lucky to avoid crashing on top of other survivors, who now all filled the surrounding area, panicking and screaming. Swimming away, to avoid being sucked under by the sinking ship, Eric somehow found Lamb among the chaos. Unable to find a space on a raft, they spent much of the night swimming from raft to raft, and away from the scene, not so much because of the risk of suction from the sinking ship dragging them down, but because they were concerned about being taken prisoners of war. That fate awaited Sub-Lieutenants Campbell, Robinson and Taylor and Leading Airman Fallon, who were captured and taken back to Germany in one of 7th Flotilla's U-boats. Thankfully, HM Ships *Convolvulus*, *Marigold* and *Deptford* chased off the German submarines before any more prisoners could be taken.

When HMS *Convolvulus*, a corvette, eventually stopped to take them aboard, Eric found the cold had made his legs unusable. 'We were supposed to climb that wet heaving cliff [the ship's side],' he recalled twenty years later. 'My legs were too weak, and I got badly skinned going up.' His CO, Lieutenant-Commander Gibson, was already aboard, as were surviving members of the squadron and some of the ship's company, who soon brought him back to life with a gulp of rum. He was extremely lucky. Out of the ship's company of 250, more than fourteen officers and ninety ratings had died.

Eric put his survival down to his physical fitness and wearing his Irving sheepskin flying jacket over his thick naval service dress of worsted jacket and trousers, flying boots and a thick jumper. He also credited the 'Mae West' flotation gear, whose unique posture kept his head above water, even when he fell asleep. Others in the water, who were less well protected against the cruel Atlantic, succumbed to the cold.

Sadly, Commander MacKendrick and Lieutenant-Commander Carline were among those lost with the ship. MacKendrick was awarded a Mention-in-Despatches and Carline was gazetted for the Distinguished Service Cross. There were also awards of the Distinguished Service Cross to Lieutenant Sleigh and Sub-Lieutenants Hutchinson, Williams and, of course, Eric.

With HMS *Audacity* lost, the remaining convoy ships steamed north, now on a direct track to reduce the time at sea, Commander Walker, the escort commander, insisted that they needed more help. Just before dawn, two Liberator long-range patrol bombers from RAF Nutts Corner, near Belfast, arrived to secure the sea ahead of the remaining ships in the convoy. After this, there was no more trouble from U-boats, as the ships steamed up the Irish Sea.

HMS *Audacity* had escorted four convoys, and its air group had flown on thirty-five days, nearly always without an alternative airfield ashore within range. The Martlets had helped sink a submarine, downed at least three (five if two written off on their return are included) Condors, scared off seven more with damage to at least three of them and engaged a Ju 88 long-range fighter. Such was *Audacity*'s undoubted success, the Ministry of War Transport no longer complained about merchant hulls being converted to aircraft carriers. It also threw its weight behind the programmes being developed in the United States to pool knowledge and experience on convoy operations. HMS *Audacity* might have sunk, but it had done its job, and then some.

The Germans were, of course, delighted that such a formidable foe had been vanquished. Vizeadmiral Karl Dönitz even named HMS *Audacity* in his end-of-year report to the Oberbefehlshaber der Kriegsmarine: 'In (convoy) HG76, the worst feature from our point of view was the presence of the aircraft carrier *Audacity*.'

Most survivors of convoy HG 76 returned to Liverpool on Christmas Eve. Others took a few days longer, as the rescue ships dispersed to unload. At least their families had been told they had survived. For other families, receiving a telegram saying 'The Admiralty regrets – missing believed killed in action' was a bitter blow just before Christmas.

When Eric marched smartly off HMS *Deptford* at Liverpool, in his pressed service dress, complete with officer's hat, he was cheered by the dockyard workers. Yet others, including the squadron boss, Donald Gibson, who had lost his naval cap and was dressed in survivors' battledress and blankets, were verbally abused, as they were thought to be German POWs. That Eric was cheered while others who were equally deserving were booed sadly rankled with many. He was already beginning to earn a reputation as someone about whom everyone had a distinct opinion.

Eric was one of five survivors of the 802 Naval Air Squadron pilots who had embarked the previous September and who was still fit for active service. Among these, he had lost close friends and shipmates, not least his squadron mates Sub-Lieutenants Norris (Pat) Patterson and Fletch.

On returning home, Eric visited Graham Fletcher's family to explain to them that their loss had not been in vain, and that Fletch had saved many lives by his prompt action against the U-boat. The Fletcher family maintained close contact with Eric, almost adopting him and certainly making him 'one of the family'. It was a relationship which continued until Eric's death in 2016, when the Fletcher family attended his memorial service at RNAS Yeovilton. Eric also visited Norris Patterson's sister Mary, and she too remained a close friend.

There is no doubt that the Battle of the Atlantic was as significant for Eric as it was for the Allied cause. Eric learned a huge amount in four months and had gained a high reputation for his deck landing skills. So much so that it was thought that his unique talent might be better suited elsewhere. Somewhere, perhaps, even more dangerous than the Battle of the Atlantic.

PART 3

Test Pilot 1942–49

Eric, seen here as a Rating Pilot, learned fighter tactics in some very weary Hurricanes discarded by the Air Ministry and passed to the Fleet Air Arm. He flew the fighter at RNAS Eastleigh and the newly opened RNAS Yeovilton in Somerset. Eric had at least one major incident when the Hurricane caught fire and he 'attracted their Lordships' displeasure' because of the number of flying accidents he had, which the Admirals thought were because of 'high jinks' rather than proficient flying.

11

Testing to the Limits

Following his exploits on HMS *Audacity*, in January 1942 Eric was granted a well-earned two-week leave. He made his way to Belfast to see Lynn, and the couple decided to take this opportunity to get married. However, as the preparations were well underway, Eric was suddenly ordered to report to the Royal Aircraft Establishment (RAE) at Farnborough. He instantly knew what this meant.

Farnborough was the cradle of British aviation and a world-leading research establishment. It was where Samuel Cody first took to the air in 1909, demonstrating to the British Army that the third dimension would be the twentieth century's new military environment, whether the generals liked it or not. It built fighters and bombers in the First World War and it was doing so once more, coming up with revolutionary machines to defeat the Luftwaffe. Often, they were experimental, but every idea, no matter how far-fetched, had the potential to be worked up into a prototype and put to the test.

To help with this, Farnborough required pilots. Flying

in wartime was dangerous but test flying was doubly dangerous. The risks taken by pilots and flight test observers to improve the capabilities of aircraft would not have normally been contemplated in peacetime. During the period 1939 to 1989, on average every year ten pilots, flight test observers or engineers were killed in test flying at the Royal Aircraft Establishments alone.[1] Unsurprisingly, only those pilots rated 'exceptional' or 'above average' were considered for such a dangerous role.

Eric's courage and skill in action aboard HMS *Audacity* had quickly established his reputation as one of the most promising pilots in the Fleet. His organized brain, small stature and tough physical form were also extremely useful attributes for a test pilot. Indeed, HMS *Audacity*'s Captain, Commander Douglas MacKendrick, had reported back to the Admiralty that Acting Sub-Lieutenant Brown's 'flair for deck landing should not be wasted'. When the Admiralty continued to receive encouraging reports of Eric's natural ability, it was clear that he might be of use to Farnborough.

If Eric was daunted by the call, he certainly didn't show it. On 5 January, in between his wedding preparations, he was flown to Farnborough, where his suitability was put to the test. He was told that he was to

1 Figures estimated by Richard Gardner at the Farnborough Air Sciences Trust.

test, and report on, the Miles Aircraft Company M20, a war emergency design from the small aircraft designer at Woodley, near Reading. Built in record time to Air Ministry specification F19/40, the latest Miles creation was a twelve-gun fighter of wooden construction, intended to be available in large numbers in the event of a German invasion in 1940–41. The new aircraft, which was powered by the latest Merlin XX engine, had a fixed undercarriage and simple handling characteristics, so that relatively inexperienced pilots could easily fly it. However, after the M20 underwent trials at the Aeroplane and Armament Experimental Establishment, Boscombe Down, in September 1940, the Royal Air Force lost interest when it was clear Spitfire production could be ramped up at the Castle Bromwich shadow factory. The immediate need for the M20 therefore evaporated.

Yet while the Royal Air Force might have lost interest, the Royal Navy had not. The Admiralty was looking for a lightweight, cost-effective plane with a disposable airframe that could be used by the Catapult Aircraft Merchantmen, not surprisingly known as CAM-ships. Rather than use a runway on a flight deck, the fighters would be launched into the air via catapult, after which the aircraft would be expected to ditch alongside – literally land in the sea and the pilot take to a dinghy and await rescue – having hopefully destroyed any marauding German bombers. With this in mind, a second prototype was built by Miles in answer to Admiralty specification N1/41 and first took to the air on 8 April 1941. The design, while still largely wood, was changed to incorporate an undercarriage that could be jettisoned for a one-way combat climb. Lord Beaverbrook, the Minister for Aircraft Production, took this a stage further and suggested that redundant Merlin III engines from Spitfires should be used. He also suggested that no undercarriage, radio nor dinghy should be fitted, and the weight saved could be taken up with a 250 lb bomb to attack submarines. It was to be a one-way ticket for the pilot.

This was the version Eric was to test at Farnborough, which was designed as a stop-gap fighter for merchant aircraft carriers and had now been fitted with a fixed undercarriage for testing, as well as an arrester hook to facilitate deck landing.

After flying the prototype M20, Eric reported that it was 'a cross between a Spitfire and Hurricane but with better range and firepower, even though the wingspan was shorter'. He liked the bubble hood, which would make a fighter pilot's job of searching the sky around him easier, but he was less sure about the proposed absence of undercarriage and general handling. Although Eric was relatively enthusiastic about the M20, the Admiralty eventually decided to drop it.

Following this test, there was no confirmation that

Eric would again be required at Farnborough, so he made his way back to Belfast, where, on 17 January 1942, he married his beloved Lynn. The *Belfast Telegraph* marked the wedding with the headline 'Scots Airman's Ulster Bride'. The wedding was officiated by the Right Reverend Dr James Woodburn, who was described at the time as a local hero, having done so much for the area, especially after the 1941 Blitz. It was also witnessed by Eric's proud father and Lynn's parents, as well as Muriel Smith, the bridesmaid, and best man Captain Alan Marsh, a Royal Marine pilot. Marsh was an instructor at RNAS Sydenham in Belfast, where he and Eric were firm, if somewhat temporary, friends. In the sad absence of Fletch, and with a best man urgently required, Marsh was happy to step in. Sadly, it seems, he and Eric lost contact completely after the war.

For Eric, the day was a happy event. Not only could he enjoy a break from hostilities and marry Lynn, but this was also a chance to reconcile himself with his difficult past as an orphan. Now, at last, he had a family he could call his own. Just as importantly, it might also prove to be helpful to his career. In the class-obsessed days of the Royal Navy, where promotion to upper ranks was still somewhat dictated by class and background, Eric knew that his status as an orphan would be frowned upon, and might even halt his progress. Marrying Lynn might therefore provide him with some respectability in this regard and prevent awkward questions being asked.

On 1 February 1942, after a brief honeymoon in Edinburgh, Eric said farewell to his bride and returned to RNAS Yeovilton and 802 Squadron. There, Eric continued building up his flying time, now standing at 346 hours 'in command'. Before the HMS *Audacity* deployment, Eric had accumulated 185 hours' day flying and five hours at night. Within a year, he had added over 800 hours to that total, to take him to the magic 1,000 hours' mark. To cap it all, the *London Gazette* of 6 March 1942 finally announced the award of his Distinguished Service Cross for 'bravery and skill in action against enemy aircraft and in protection of a convoy against heavy and sustained enemy attacks'.

Five days later, Eric attended Buckingham Palace for the investiture. Joining him were Lynn and her parents, as well as Eric's father. It would not be the last time he walked the red carpet to the monarch for an award. After the investiture, made personally by HM King George VI, the party went for lunch, but there was no time for any prolonged celebration, as Eric returned to Yeovilton that same afternoon.

Still on the books of 802 Squadron, Eric's next task was training for catapult use from an aircraft carrier's deck. Until now, Eric's flying had been in the Martlet Mk I and Mk III, which had used free take-off techniques, taking as much of HMS *Audacity*'s flight deck as possible to get airborne, usually less than 300 feet. However, 802 Squadron had now been earmarked for a carrier with a catapult system. Despite Eric's hundreds of hours of flying time, and take-offs, this was to be something very different.

At HMS *Daedalus*, RNAS Lee-on-Solent, Eric and his squadron mate, Sub-Lieutenant Graham Hutchinson, were due to fly a Swordfish, which would be launched by a Salmon Telescopic Tube Catapult. While Eric had previously flown a Swordfish, he recalled that Hutchinson had not. Both were given a quick briefing and a walk-round of the aircraft on the concrete before it was craned on to the catapult, which faced out to the Solent.

The pilots climbed in, with one in the pilot's seat and the other in the observer's position behind. Pulling the seat straps tight, Eric started the Pegasus engine, then, once it was at full power, gave the thumbs-up. At this moment, the piston in the catapult tube, which was held at pressure, was released, launching the mounted Swordfish into the air at a flying speed of about 70 mph. It was an exhilarating experience, and one that gave Eric confidence to test catapults ashore, and afloat, during the rest of the war.

On 18 April 1942, Eric joined his messmates from 802 Squadron at RNAS St Merryn in Cornwall for three weeks of extensive flying practice, in preparation for a Mediterranean deployment aboard the escort carrier HMS *Avenger*. Once more, Eric continued to turn heads with his ability. A note in his confidential report from Lieutenant-Commander Donald Gibson, his squadron boss, rated his flying skill as 'above average' and entered a note in his war training scheme register as 'a pilot of exceptional ability who has done very well indeed. Takes a very keen interest in his work both in the air and on the ground. Tactful and loyal.' This seems to have allowed his seniority, lost in 1940 after the youthful high jinks with Fletch beating up airfields, to be restored.

After all of this training Eric was eager to return to action at sea, but the Admiralty had other plans. Things at Farnborough had clearly gone well, and he was instead told that he would be kept behind for classified trials. To Eric, this was somewhat of a blow. In his eyes, protecting convoys, and being on a carrier, was where the serious action was to be found. This was also where his squadron would be deployed, and he now felt very much one of the gang. Despite this, there were orders to follow, and Eric had no choice but to remain in the UK as 802 returned to sea. Dismayed at this turn of events, he didn't yet realize that the next few years would be the most exciting, and dangerous, of his career.

Behind these classified trials was the important issue of reinforcing Malta, at the time described as the most bombed place on earth. It was vital that fighters were sent to Gibraltar on aircraft carriers as a matter of urgency, from which they would fly to defend Malta. However, there was now a limited supply of Martlets coming from America, while there were also concerns that the fighter was not entirely suitable for air carrier combat, as it did not have self-sealing fuel tanks or adequate pilot armour. In response, the Admiralty initially asked the Royal Air Force for any surplus Spitfires, which they would try to convert for naval use. The RAF refused, claiming that at this time the Spitfire was vital to homeland defence, and none could be spared. They did, however, offer to provide the Admiralty with some used Hurricanes. These were far from ideal. The Hurricanes had mostly served in the Battle of Britain, and while they had since been through the Civilian Repair Organization at Cowley, near Oxford, they were really showing their age, as well as their lack of suitability for front-line service. Indeed, Eric described them as 'beaten up and very teased'. From bitter experience, he also knew they were unreliable. For example, while Eric was flying over Somerset with 802 Squadron, his Hurricane 'started to sizzle'. Thankfully, he managed to land at Yeovilton before it caught fire.

While these Hurricanes were old and somewhat battered, they were also not created with naval use in mind. The Admiralty merely stuck an arrester hook on them, and re-designated them Sea Hurricane Mk Ibs. Still, there was another troubling question for the Admiralty to answer: could the average pilot handle a Sea Hurricane take-off from a short, narrow and probably pitching deck?

Eric wasn't so sure. When at 802 Squadron, he had built up many hours on the Sea Hurricane Mk Ib variant with its arrester hook. He knew the hook made the machine tail-heavy in the air and difficult to fly. Because of the Hurricane's long nose, this also made the sight line difficult on approach to small aircraft carriers. It was therefore a challenge not only to land it on a small deck, but then to catch the flight deck arrester wires. Despite Eric's misgivings, there was no choice. Somehow, he had to master Hurricane landings so other pilots could do the same.

With this in mind, on 11 May, Eric headed to Scotland for three weeks of intensive training, concentrating on Hurricane landing trials aboard HMS *Avenger*. As part of 768 Naval Air Squadron, he would be based at RNAS Crail (Arbroath), also known as HMS *Jackdaw*. Situated on a promontory, south of St Andrews in Fife, it was poised as if about to leap into the North Sea. Airfields tend to be sited where the wind blows, but Crail is especially vulnerable to strong winds, both challenging and exciting for a naval aviator. Making matters even more difficult for Eric at this time, the weather was awful, with the wind howling, and visibility down to almost zero.

This was all extremely dangerous, and even Eric admitted that he was afraid, often saying that 'only fools know no fear'. To counter this, he said that he always prepared for the worst and was therefore able to control his emotions should things not go to plan. He visualized every single issue that might occur and had an alternative plan in mind. Thanks to this mindset, Eric found that with reasonable care, he was able to land the Hurricane with little problem, although he did have to utilize a straight approach, to prevent the nose becoming too heavy, which meant his view was somewhat obscured.

There was no doubt that the Hurricane was far from ideal for a carrier, and had only enough fuel to sustain itself for one hour at combat power, compared to the Martlet's two hours. Still, Eric reasoned it would certainly do for now. Indeed, soon after, the Sea Hurricanes played an important role in the battle to save Malta, especially during 'Operation Pedestal', where they assisted operations to deliver enough supplies for the population, and military forces, to resist invasion.

Having proved his worth landing a Hurricane, the Admiralty had another, even more dangerous, job in mind for Eric. He was subsequently posted to HMS *Argus*, an old training ship, where he was to serve as a deck landing instructor to new naval pilots. If he thought landing a Hurricane on a carrier was frightening, then this was a different matter altogether.

During training, Eric would sit in the back of a Fairey Fulmar or Fairey Swordfish, without the benefit of dual controls, while in the front cockpit a young pilot took control of the aircraft. Eric remarked that this was 'like playing Russian roulette'. Which was to say the least. He was putting his life in the hands of another pilot, in what was considered a very challenging landing.

In an attempt to calm the pilot, as well as himself, Eric always ensured that they went through his twelve-point principles before landing. The most important ones to understand were the 'numbers' – the landing speed, stall speed and speed not to exceed – the weights and centre of gravity and, most important of all, understanding the means of escape in the event of a problem. It was only after mentally checking through this list, and being prepared for the worst, that Eric signalled for the pilot to land.

Things did not always run smoothly. On one occasion, a student pilot 'gunned' the engine with the force of the landing, causing the aircraft to snake across the flight deck and tip over the side of the carrier, before being caught by an arrester wire. Eric, the student and the Fulmar were left dangling in space, just above the waves, 'with the propeller scratching the ocean'. Eric was understandably exultant when the single-seat Sea

Hurricane replaced the Fulmar and there was no room for the instructor in the back. He admitted that he was glad to see the back of this 'highly nerve raising job'.

Following this, Eric was ordered to go to RAF North Weald, to fly the twin-engined Westland Whirlwind[2] – 'a great fighter which might have been so good with the right engine' – that had been tested by Westland test pilot Harald Penrose. Penrose was the test pilot Eric admired most and the one who inspired him to develop his own distinct and enduring set of rules for a test pilot. These sage words of advice were captured in Penrose's autobiography, *No Echo in the Sky*: 'Flying must always be made in such a manner that the unknown can be met in a way that leaves some avenue of escape. As a last resort, there is the parachute: sometimes fate trembles on the finest thread.'

It was invaluable advice as, during this period, Eric met his old nemesis again. On 27 July 1942, Eric found himself flying in a Martlet, over the North Channel, from Machrihanish to Northern Ireland. Yet what was intended to be a quick overnight in Belfast to see Lynn turned into a classic chase at 50 feet across the Hebrides. The reason for the change of plan was the sudden appearance of a German Condor bomber coming from

2 The Whirlwind had been designed as a Spitfire replacement, with two less than adequate Rolls-Royce Peregrine engines and four nose-mounted cannon.

the direction of Eire. Outside gunnery range, the Condor was circling three destroyers below that were presumably awaiting a convoy in the Clyde approaches. Eric had no option but to engage if the convoy was to remain intact.

In usual circumstances, the Condor, with its prodigious defensive armament of cannon and machine guns, was a formidable foe. However, this situation was far more dangerous than the norm. Eric's Martlet was unarmed, a standard procedure until then for trials' aircraft. All Eric could do was tail the bomber, dodging its defensive fire, until fighters could be scrambled to tackle it.

Swinging north, the Condor headed directly for the Outer Hebrides, probably hoping to lose the irritating Martlet in the Atlantic as the fighter's fuel reserves dwindled. The Martlet, of course, was designed as a naval fighter for the Americans, with operations over the Pacific in mind, so it had at least four hours of endurance. Eric would need every last drop of fuel, because it would take him until Colonsay, that beautiful island in the Inner Hebrides between Islay and Mull, before the Condor came close to being caught at 200 mph.

Even then, Eric's calls to the local air defence controller had only just been answered, with two Beaufighters hastily scrambled. Until they arrived, it was vital that Eric kept contact with the Condor. But it was becoming harder to keep dodging the hail of fire erupting from it, and Eric was also becoming increasingly nervous about his dwindling fuel supplies. When it reached 'bingo', he had no choice but to break off the pursuit over Barra in the Outer Hebrides and return to the nearest airfield, landing at Crossapol on the Isle of Tiree for fuel and, as it was nearly dark, accommodation.

On landing, he noticed that the German gunners had been accurate, and his Martlet had holes in the tail empennage. It had been an enormously brave pursuit and one that might well have saved many lives. Following Eric's alert, the Beaufighters claimed to have shot the Condor down over Benbecula. While German records found after the war did not confirm the loss of a Condor that day, it may be that the Condor was able to limp on to Norway. Whatever the end result, the unarmed Eric, and the Beaufighters, had kept the Condor occupied and away from the British convoy.

With Eric's reputation as a fearless and thorough test pilot growing, the Admiralty once again thought it might make better use of his talents. As such, his greatest challenge yet lay in store.

Eric launches from the deck of the trials carrier HMS *Pretoria Castle* in July 1945, flying Seafire 15, his Seafire TM379, the prototype F45.

12

The Seafire

On 1 August 1942, Eric was posted to HMS *Blackcap*, the Royal Naval Air Station at Stretton in Cheshire. Appointed to Senior Pilot, he joined the newly formed 897 Naval Air Squadron, and while he had previously mastered landing the Sea Hurricane, he now had a more problematic test ahead.

The Air Ministry had finally agreed to provide the Admiralty with Spitfires for use on carriers. *Blackcap* was subsequently equipped with the newly arrived Seafire Mk Ib fighters,[1] which were little more than hooked Spitfires. Lieutenant-Commander Hugh Bramwell DSO DSC was the first to take the Seafire to sea, for a series of tests on HMS *Illustrious*. He found that the classic fighter was not easy to handle, and because of the Seafire's long nose, which obscured his view, he needed to curve the approach to the deck, which allowed him to see it until the very last moment. The biggest concern, however,

1 This Seafire was a modified Spitfire Mk Vb, a cannon-armed fighter already in service with Fighter Command.

was the Seafire's delicate undercarriage, which was not as robust as the Hurricane's. Fleet Air Arm pilots, especially when landing back in poor weather had a habit, through necessity, of making heavy landings. Bramwell was not sure the Seafire's undercarriage could repeatedly sustain such an impact. On the plus side, if the Seafire could prove to be operational at sea, it promised to be the high-performance interceptor for which the Admiralty was so desperate.

Eric was therefore one of a team of test pilots tasked with developing a way for the Seafire to land more safely on deck. This was not a job for the faint-hearted, but it was integral to the war effort. If the test pilots couldn't work this out and train pilots to land the Seafire successfully, then convoys would be at risk. Yet even Eric recognized the difficulties: 'Everyone admired the Spitfire and itched to fly it – but from an aircraft carrier? That was a horse of a very different colour!'

He practised on dummy decks relentlessly, but these were no substitute for the real thing, particularly as the aircraft carrier selected for the Seafire embarkation was HMS *Biter.* Commissioned into the Royal Navy in May 1942, *Biter* was another merchant ship converted into an aircraft carrier and, like HMS *Audacity*, boasted a short deck of just 492 feet. However, as she was converted in New York, she was equipped with the latest US naval flight deck technology, including the American Tail

Down Catapult for launching aircraft. This was very different from the British version, thankfully in a good way.

The British method involved hoisting the machine up on a trolley, where four claws connected up with four spools on the belly of the aeroplane. When the trigger was pulled, trolley and aircraft shot down the deck, with the trolley fetching up against a stop near the bows, its front legs collapsed forwards, while the aeroplane was propelled into the air. In contrast, the American system had a shuttle, which ran along a slot in the deck for'ard. At the after end of the slot was a hold-back hook. The aircraft was simply taxied over the slot, its belly fixed loosely to the shuttle by a strop, its tail to the hold-back, which incorporated a breaking ring in it, fixed to break at a certain pressure. With the aircraft in this position the shuttle was then tensioned till the strop and hold-back hook were taut. When the shuttle was fired, the hook broke, and the aeroplane was accelerated into the air and off the deck.

When testing take-offs from HMS *Biter*, Eric was slightly apprehensive at doing so in the new American Tail Down Catapult. He needn't have worried. He found not only that it took less time to prepare, but that he also felt far more 'g' force on launch, and he gave it full marks. Take-off in Seafires from *Biter* was therefore not an issue. What really concerned the Admiralty, and Eric, was the landing.

The deck, like HMS *Audacity*'s, was short, but Eric had already proven that this would not be of too much concern. The greatest threat was the Seafire's long nose, which obscured the view, as well as the ability of the fighter's delicate undercarriage to sustain the landing. As Eric said, 'There was a certain air of fragility about the aeroplane; a ballerina-like delicacy that seemed inconsistent with the demanding, muscle-taxing scenario of shipboard operations.'

On 11 September, Eric took off from Machrihanish and made his way to HMS *Biter* for his first attempt at landing a Seafire on deck. Concentrating intently, he decided to utilize the 'Bramwell crab',[2] using a curving, sharply banked approach to the carrier, as the Service Trials Unit (STU) Commander, Hugh Bramwell, had advised. This at least allowed him to see the deck until the last moment, but now came the toughest part of all: putting the fighter down safely. Hitting the deck as gently as possible, the Seafire caught an arrester wire and came to a stop with the undercarriage still intact. Eric breathed a sigh of relief. He deemed the landing a total success. In fact, it had all seemed very straightforward,

2 The Bramwell crab technique was developed to give the pilot a constant view of the flight deck of an aircraft carrier. The approach was made to the deck in a wide curve from the downwind leg with a short final approach on to the deck.

and he thought he had been worried over nothing. But then something horrified him.

Looking up at the ship's 'island', he now noticed flag G – the 'go away' signal – was flying. He also realized that no one was on deck or had come out to congratulate him. Such was his concentration on the job in hand that he hadn't realized it was lunchtime, and there was not even a deck landing officer present. To his horror, he became aware that not only was the ship not prepared for him to land, but the wire he had somehow caught was not even properly rigged. It was a miracle he had caught it. To compound matters, the flag showed the ship was crosswind by 25 degrees, which would have made any landing even more difficult. Eric was decidedly shamefaced. 'I felt very small,' he admitted, 'very stupid, inexcusably careless and stupendously lucky.' On the other hand, HMS *Biter*'s Captain, Abel Smith, was impressed. If Eric could land a Seafire when the ship was out of wind and not rigged to receive aircraft, then as far as he was concerned the trials were a complete success. He subsequently made a signal ashore to that effect.

Just to be sure, and this time with HMS *Biter* expecting him, Eric decided to make one more landing attempt, this time into wind. It seems he almost pushed his luck too far. On this occasion he approached the deck too fast, broke off the Seafire's arrester hook and could only stop by running the fighter into the ship's superstructure.

'It was a major black,' Eric wrote in his logbook. Nevertheless, Captain Smith's signal ashore stood, even if the damaged Seafire had to be craned off the ship at Greenock. Perhaps this should have been a sign that all was not as safe with the Seafire as was first thought.

Once cleared for service, in November 1942 HMS *Biter*, and the Seafire, were deployed to cover the Allied landings in North Africa, code-named Operation Torch. However, while there was still no choice but to use it, the Seafire struggled at sea, with frequent engine and airframe problems, not to mention landing mishaps. Although Eric had been able to land the Seafire, the average pilot found it enormously difficult, mostly due to the fact that the fighter was simply not robust enough for air carrier landings, with forty-eight Seafires out of eighty written off in non-combat flying, mainly landing accidents. Still, in adverse circumstances, it was better to have it than not, and it played its part in a successful operation. Indeed, until the last days of the war, the Seafire was used to provide air cover for the Allied invasion of Sicily and then Italy, while in 1944 it supported the Allied ground forces during the Normandy landings and Operation Dragoon in southern France.

However, as Eric rubbed shoulders with those in the upper echelons of the Royal Navy, he was becoming ever more conscious that he was from a different class, upbringing and background. He soon realized that while there might have been an influx of citizen sailors through wartime expedience, there was still an invisible bar to reaching the top for those without a grounding at the Britannia Royal Naval College at Dartmouth. Not only had Eric not attended Dartmouth, he was also keeping a secret regarding his adoption. He realized that together this would surely be a substantial barrier to him advancing. Nevertheless, Eric was keen to break through that glass ceiling and move upwards in the service, which he had come to love. For this to be possible, he thought that his test flying abilities would be the answer.

With this in mind, Eric threw himself into flying at the Service Trials Unit. It became an obsession. He ignored weekends and games afternoons, other than the occasional game of rugby, and did all he could to prove that he belonged in this supposed class of company. During this period of intense flying, Eric completed deck landings and launches on twenty-three different aircraft carriers, mainly newly commissioned Escort types and the Colossus-class Light Fleet carriers, which were being built to an Admiralty design, but in yards with little or no naval experience. He took a variety of aeroplanes to these ships for their Sea Acceptance Trials (Air), including the Swordfish, Albacore, Fulmar, Sea Hurricane, Seafire, the new American Avenger (which the British insisted on calling the Tarpon), Barracuda and the Wildcat (still at that time called the Martlet).

Even free Saturday afternoons soon became scarce, as Eric enjoyed spending time with the junior 'boffins' – mainly aeronautical engineering apprentices – going over the latest inventions and refinements. 'He seemed to prefer talking to us rather than going to the bar with the other pilots,' remembers Flight Test Engineering Apprentice Geoffrey Cooper. 'We would talk through problems and get his insight.'

Such was Eric's fixation and need to remember all the technicalities of the aircraft he flew, he kept a diary, meticulously recording the layouts of the cockpits, fuel systems, hydraulic systems, engine settings, landing speeds and other data. There was little time to make any real friendships, and Eric wasn't interested in any event. He found it easier to focus on the machines than any human beings. This was emphasized when Geoffrey Cooper joined Eric as 'ballast' for a catapult launch of an Auster light observation aircraft. Cooper recalls that Eric did not want 'talking ballast', and 'every time I started to speak, Eric advanced the throttle to drown my speech; when I stopped talking, he slackened it off. I knew my place.'

This obsession and effort had some success, as Eric's confidential report for September 1942 to January 1943 records that he was 'an exceptional pilot with plenty of experience who shows a keen appreciation of his work. Of an industrious nature and phlegmatic disposition, he should prove a very good flight leader in a fighter Squadron.' It is signed by Captain R. M. T. Taylor, who also recommended Eric for accelerated promotion to Lieutenant. This resulted in Eric being promoted to Temporary Lieutenant (A) on 1 April 1943, by which time the STU had moved to RNAS Crail, the Fife aerodrome that Eric knew well.

Besides testing new aeroplanes, equipment and techniques, the unit was also the centre for the Royal Navy's evaluation of captured enemy types. This was perhaps the most dangerous and exciting job of all for a test pilot. There was often no instruction manual or guidance. All a test pilot could do was to take the captured machine into the air and work it out from there. Indeed, this was also one of the most important jobs in the war. Test pilots like Eric were tasked with understanding the strengths and weaknesses of these captured enemy aircraft, with their findings filtered down to the battlefield. Often this information would help save lives and win battles.

The first machine Eric took aloft was the nippy Italian Fiat CR 42 fighter, which had been captured in the Western Desert a year earlier. Things were at least made slightly easier as Eric had flown the British equivalent, the Gloster Gladiator fighter, during his fighter pilot training in September 1940. In his subsequent report Eric commented that the Italian fighter had a slight advantage over the Gladiator in most respects, including speed, manoeuvrability and control harmony, but not in firepower, or comfort to fly, as the CR 42 had an open cockpit. However, in comparing the Italian machine to the Hawker Hurricane, Eric was most emphatic: 'It is not in the same class except the CR 42 can out-turn the Hurricane and it is nicer to handle within its speed range.'

Soon Eric would have his pick of Italian aeroplanes to fly, as in September 1943 Italy was invaded and surrendered to the Allies, with all Italian aircraft made available to the STU for evaluation. While Italy had fallen, tests still needed to be conducted, as Italian Republican Fascists, still loyal to the fallen dictator Benito Mussolini, continued to fight.

The new force was called the Aeronautica Nazionale Repubblicana and included the Fiat G55 and Macchi 205V in its inventory. The latter, Eric said, 'was on a virtual par with the Spitfire Mk IX and Mustang Mk III [P-51B/C in US service]', which were then the standard day fighters of the Royal Air Force and US Army Air Force respectively. But for Eric, an important flying trial was to see the capabilities of the Cant Z1007, which had shot down one of his 802 Squadron messmates after Eric had been posted to the Service Trials Unit. He said of the Cant Z1007, 'It was quite well armed, comfortable to fly and was a good, stable weapons' platform.'

By the turn of 1942, Eric had logged 660 hours' air time, but still he would not rest. In the first six months of 1943 he somehow added another incredible 800 hours, an astonishing number. Amongst duties assigned to him over New Year was deck landing instruction to a front-line Spitfire unit, No. 65 (Fighter) Squadron. After initial aerodrome dummy deck landings (ADDLs), the pilots were to attempt the real thing aboard HMS *Argus* in the Clyde exercise area. The pace of work was dictated by the need to have pilots trained, and systems proved, for the opening of the eventual second front in Normandy.

While Eric was an exceptional trainer, and worked his students hard, it was never something he particularly enjoyed. He'd much rather be in the air himself, either fighting or testing new and exciting aircraft. As usual, whenever Eric felt boredom start to creep in, he found himself in trouble.

Flying out of Arbroath on 3 March, en route to Crail, just across the water, Eric decided to show off and buzz the control tower. In doing so he caused mayhem by carrying away the radio mast and aerials on the wing of his Martlet. 'Another major black,' he recorded in his logbook.

As well as this there were also multiple incidents concerning bridges, something Eric was never able to resist. Despite it being dangerous, and prohibited, he relished the challenge of flying a high-performance aeroplane through or under their spans. Indeed, after escort carrier deck landing trials in the Firth of Forth, Eric couldn't resist looping through all three spans of the Forth Bridge. However, someone reported his antics to the police. Thankfully, it was the Royal Air Force who were accused, and no one thought to come looking for the culprit in the Royal Navy. Eric was lucky not to get into serious trouble. Despite this, he remained undeterred from repeating this feat, next time taking a Seafire through the centre span of the Tay Bridge, followed by a Hurricane and a Swordfish, as well as a Vought Kingfisher floatplane.

Eric clearly needed a boost of adrenaline and was looking for a new challenge. Soon he would get what he was after, as one of the most infamous squadrons of the war put him to the test, in more ways than one.

Eric wasn't fearless, but he was a practitioner of managing risk. When asked to teach deck landing techniques to Canadian fighter pilots, he found himself volunteered to fly the Spitfire on operational sorties over occupied France and even engaged in combat against German fighters. Just what his bosses at Farnborough would have thought if their Chief Naval Test Pilot had been shot down was never tested, and Eric didn't let on until after the war.

13
The Canadians

While Eric was with the STU in Scotland, a signal reached him ordering him to fly south and report to the newly formed Canadian Fighter Wing. The Wing had recently taken over RAF Kenley and nearby RAF Redhill for fighter operations over enemy territory. The Canadians were regarded as tough, daring and audacious pilots, if rather ill-disciplined on the ground, especially in the vicinity of the local hostelries. However, they had acquitted themselves well against the enemy, most notably during the Battle of Britain, where they became known as the 'Wolfpack' because of their aggressive skills at hunting down the enemy.

Such was their reputation that when Headquarters Fighter Command, at RAF Bentley Priory, decided that some of its best pilots might be needed to support Allied landings at Salerno in Italy, the Canadians were an obvious choice. However, the lack of suitable airfields near Italy meant that they would have to operate off aircraft carriers. This was something of which they had no experience. Eric was therefore ordered to teach them how to land Spitfires on a purpose-built 'dummy deck', painted on the runway at RAF Kenley. The Canadians were less than enthused. Before they would even contemplate such 'a damned silly thing' as attempting a deck landing, they wanted to test the mettle of the man who had been sent to teach them.

A typical sortie for the Canadians was a 'rhubarb' operation, where at times of low cloud and poor visibility, fighter or fighter-bomber sections would cross the English Channel and search for opportunity targets, such as railway locomotives and rolling stock, aircraft on the ground, enemy troops and vehicles on roads. However, the Canadians' next mission was a 'circus' operation, which involved a bombing raid on Caen-Carpiquet airfield in Normandy by 24 Lockheed Ventura light bombers, based at RAF Tangmere on the Sussex coast, accompanied by large escorts to tempt enemy fighters into the air. To put their instructor to the ultimate test, Eric was invited to join them. He could hardly say no. Besides, this was just the excitement for which he was looking.

In the early morning of 4 April 1943, Eric and the Canadians set off from Redhill in a gaggle of Spitfires, consisting of two squadrons: No. 411 (City of York) and No. 421 (the Red Indians). Eric was delighted to be at the controls of an operational Spitfire, even the now slightly outdated Mk Vb. After rendezvousing with the

Lockheed Ventura bombers at Selsey Bill, south-west of Redhill, Eric and the Canadian fighters climbed to 18,000 feet over the Channel to provide top cover.

As the fighters flew towards Occupied Territory, Eric felt at home in the rather cramped Spitfire cockpit. It was as if the Spitfire design team had guessed Eric's stature and made the cockpit hood just the right fit for him. He adjusted the seat, in order to take a wide view of the sky, and then dropped it again, to present as small a profile as possible against any enemy fighter's cannon shells. He had been caught by a German shell before and he was determined it would not happen again.

For the first time in a while Eric had butterflies in his stomach. He was carrying in his head secrets of test flying of most Allied aircraft. If he was captured, he would be a treasure trove of information for the enemy. It is surprising that he was allowed to venture across the Channel on such a mission, but perhaps 'the powers that be' did not actually know what was planned. It would be typical of Eric to accept the risk without checking first.

Down below, the medieval city of Caen was spread out like a map. Eric was not to realize that this was the last time he would see William the Conqueror's capital intact. Within fifteen months, soon after D-Day, the Second Tactical Air Force and the combined might of Bomber Command and the USAAF would 'put the city in the street', as General Patton liked to say. For now, it was a splendid sight, with the airfield itself, about 5 miles to the west of the city centre, only discharging light flak. The reason for which soon became clear.

Rising in response to the bombers and fighters sweeping in from the sea, a complete *Jagdgeschwader* of mighty Focke-Wulf Fw 190 fighters appeared in the distance. The supporting yellow-tailed Fw 190s of JG2 Richthofen, from nearby Beaumont-le-Roger, were also gathering. A short, sharp battle erupted, with the Allied aeroplanes turning for home after claiming a handful of German fighters without any loss. Eric was intoxicated. 'This kind of flying was a revelation to me,' he said. 'An incomparably satisfying business, in which a pilot lived on his wits and reflexes all the time.' But this was not to be the end of the day's combat. Far from it. After a quick refuel, Eric and the Canadians were back in the air and again heading across the Channel.

Eric's second sortie that day was with Nos 403 (City of Calgary) and 416 (City of Oshawa) Squadrons from the Kenley Wing. The target for this daylight raid was the Renault Motor Works in the Parisian suburbs of Boulogne-Billancourt, where the company was producing 18,000 army trucks a year for the Wehrmacht's campaign in Russia.

The Canadians had established a reputation for daring low-level strikes on enemy transport on roads or in the Channel. But on this sortie, for the rendezvous at

Beachy Head, Eric joined the wing flying the high escort, in the newly introduced Spitfire Mk IX, with ninety-four American B-17G Flying Fortresses in their care.

As the Spitfires swept on, past the turning point at Quiberville on the Upper Normandy coast, just 5 miles west of Dieppe, the Renault Motor Works came into sight. A raging dogfight quickly developed, with the Canadians shooting down five of the new Focke-Wulf Fw 190 fighters, but losing four US bombers and seven Allied fighters. 'Hell of a scrap with Fw 190s,' Eric recorded later. 'Lord, I was frightened.' That didn't stop him from engaging with a yellow-tailed Fw 190A over Rouen and having the satisfaction of later seeing camera-gun film footage, which showed he achieved cannon strikes on the German fighter. As a result, Eric's count of enemy aircraft received the addition of a 'damaged'.

After an hour and forty minutes, the American bombers began to extricate themselves from the target area, with their mission complete. On landing back at Kenley, there was enough time for a cup of sweet tea and a jam sandwich while the fighters were rearmed, and then they were off again. The third and final sortie of the day to the French coast was the shortest. With Eric in the leading section, the Canadians were airborne for just an hour and fifteen minutes.

On the Tuesday, Eric returned to the air once more. This time the operation was a straightforward fighter sweep of the French coast, called a *rodeo* mission. The Canadians provided the high-level escort, and by now Eric was getting used to the adrenaline rush from seeing the enemy coast and the anticipation of a dogfight. But today, there was no dogfight, and the fighters returned to Kenley, their gun-port tape still intact, signifying that the weapons had not been fired.

By Wednesday, the Canadians were starting to get to know and trust Eric. Taking a Seafire Mk IIc down from Kenley to Redhill, he joined No. 411 Squadron for a planning and briefing session prior to the next day's free-ranging rhubarb across the French countryside from Pas-de-Calais to the Seine estuary and then home to Redhill. This was somewhat of a free-for-all, and Eric fired his cannon at a signal box, two army trucks and a barge on one of the many canals used by the German construction teams to bring material to the Atlantic Wall. On the way back across the Channel, several fast enemy motorboats – R boats and E boats probably – were also shot up. Fire from these, which was completely ineffective, was the only opposition faced by the Canadians that day. Eric had thoroughly enjoyed himself and called it all 'wizard fun'.

Having spent about six hours flying with the Canadians, Eric had more than proven his worth. He was thus finally 'allowed' to demonstrate the dummy deck landing techniques at Kenley, and then at RAF Digby in Lincolnshire. Despite this, the Canadians were not required in Salerno and therefore did not put their deck landing skills to the test.

After so much excitement, Eric returned to Crail for the spring of 1943. Unsurprisingly, boredom soon struck again. Even flying didn't set the pulse racing. During radio trials, Eric was cold and fed up as he flew around in circles for up to three hours at a stretch while others tested the equipment. This was especially true when flying the Swordfish, as its open cockpit saw him pummelled by the wind and rain. The most excitement came during a so-called 'jinx party', a term Eric used for describing an aircraft incident.

During a radio altimeter test, his Barracuda lost power near Dunino airfield, 5 miles north-west of Crail. Eric had no choice but to find a stretch of open country to 'dead stick' the big torpedo-bomber. After sending out a distress call, he set the Barracuda to glide, with the large 'barn door' flaps deployed for lift rather than drag, then turned off the switches to lessen the risk of electrical fire. With this, he tightened his harness and looked below for a field where he might be able to bring the Barracuda down.

However, the land around Dunino airfield had been carefully prepared to repel landings, rather than welcome them. Invasion obstacles, such as poles and wires, were still spread across the landscape from the time when the Free Polish Army had been in residence on anti-invasion duties in 1940–41. With little option, Eric had to bring the aeroplane down, recording that the crash landing's sequence was: 'Flew through telephone wires and hit a tree with one wing; bounced about; mowed down the anti-invasion poles, two brick walls and a small hillock. A really shaky do.' Luckily, the Barracuda was quite a bulldozer, and its shape and size meant that he wasn't hurt. The aeroplane was, however, a write-off.

While 'jinx parties' were as exciting as it got at this time, entertainment away from the airfield was also hard to come by. The nearest excitement was to be found in St Andrews, a bus ride away, and nothing exciting happened there on a Sunday. The only things to look forward to were weekend film nights in the wardroom and the ENSA (Entertainments National Service Association), which brought film, radio and stage stars to outlying military establishments.

Opportunities to see Lynn at this time were also few and far between. As she had once dreamed, Lynn was using her talent to entertain the troops, singing at bases across the country and regularly appearing on BBC Radio. Whenever there was a chance they could meet, Eric would 'borrow' an aeroplane and make his way to Belfast. On one occasion, he even managed to ferry her across to Scotland in the back of a Firefly fighter, which was probably her first flight, as recorded in the logbook which Eric created for her.

Thankfully, Eric's two years at the STU were soon up. He was subsequently posted to Boscombe Down in Wiltshire, where in-service aircraft were developed into war-fighting machines. This work would be more about weapons, and weapons delivery, rather than pure flying. It was also the next step on the test pilot's ladder to the top, with Eric posted as naval test pilot in December 1943.

However, in early January 1944, Eric received cryptic orders for a top-secret mission. He was to board a Lancaster at Prestwick and make his way to southern Italy, which had recently been liberated by the Allies. Finally, some real excitement.

When Eric arrived in southern Italy, his mission soon became clear. He was directed to carry out handling tests on captured Italian aircraft, to assist the continued fight in the Mediterranean theatre of operations. As part of an Enemy Aircraft Assessment team, Eric flew the Caproni Ca 309 twin-engined reconnaissance bomber, which had been used as a transport since 1940, and the Ca 311 light bomber, of a similar size. These were both lacklustre aeroplanes, and Eric found he could quickly adapt to a second engine.

The most joyful part of the two-day trip was at Brindisi, where he had an hour in the Macchi C 202

Folgore (Thunderbolt), powered by a German-made Daimler-Benz engine. The performance of both airframe and engine resulted in Eric's high praise for the extremely manoeuvrable fighter pilot's aeroplane. Most importantly, Eric's test results were filtered to Allied pilots, who could then use them against the enemy, as well as to Farnborough, where anything unusual might prove to be useful for Britain's own aircraft.

The trip to Italy was short and sweet, but after returning to Boscombe Down on 9 January, there was no time for Eric to rest on his laurels. The Admiralty had already lined up another mission for him. For once, Eric thought it might be a step too far.

Eric was very proud of his entries in the *Guinness Book of Records*, including the first ever landing of a twin-engined aeroplane on an aircraft carrier on 25 March 1944. It wasn't just that a record was needed to be set but rather that there was an operational concept to be proven.

14
The Mosquito

A few days after Eric's return from Italy, the telephone rang in his Boscombe Down office. The voice at the other end identified himself as Wing Commander Roland 'Roly' Falk, the Chief Test Pilot at Farnborough.

'Have you ever flown a Mosquito?' Falk asked.

'No, sir,' Eric answered, although he had flown similar multi-engined-type aircraft in recent weeks, including the American-built Martin Baltimore, the old Bristol Bombay transport and two bomber types which had now become obsolescent: the twin-engined Manchester (forerunner of the Lancaster) and the four-engined Short Stirling.

'We want you to put a Mosquito on to a carrier deck by March,' Falk continued. 'Can you do it?'

Eric took a deep breath. No one had ever put a twin-engined aeroplane on a deck before. Until now, the heaviest aeroplane to fly on to a British aircraft carrier was the Grumman Avenger, a big brute of 10,845 lbs landing weight. At 21,000 lbs, an operational Mosquito Mk VI was twice as heavy. The Mosquito was also constructed with wood, which many experts predicted would break apart on landing upon a carrier. Moreover, the wingspan was 54 feet, which would bring it uncomfortably close to the carrier's island.

In spite of all of these negatives, Eric was tempted. The opportunity to be the first to achieve such a feat, and at Farnborough no less, was extremely enticing for an ambitious young pilot. If he could succeed, this might just smash the class barriers in the Royal Navy and see him move up the ranks at speed. With an appetite for excitement, and 'the rash bravado of youth', Eric replied, 'I'm sure it can be done, sir.' But this would be far more difficult than Eric, as skilled a pilot as he was, had envisaged.

Eric was not actually the first choice to test the Mosquito in this fashion. That honour initially fell to Temporary Lieutenant (A) Kenneth Robertson DSC, another volunteer reservist. However, on 14 January 1944, while Robertson was testing the Seafire on HMS *Chaser* in the Clyde estuary, the rocket-assisted take-off gear failed at a critical moment, making the Seafire violently swing off the deck. Robertson was lost at sea in the subsequent crash. With the prospect of replacing Robertson at Farnborough, Eric knew what he was getting into. 'I went there in dead man's shoes,' he said.

Things were made even more difficult when he was informed he had just two months to master the trials.

This was a very tight turnaround. He also didn't initially know why he was being asked to deck-land a Mosquito, or for what special operation such skills would be required. The answer would soon become clear: Operation Highball.

Just like the 'Dambusters', it was planned that a Mosquito would drop a bouncing bomb, which would spin at low level and skip across any defences, to hit, and hopefully sink, the target, in this case the fearsome German battleship *Tirpitz*. Dubbed 'the Beast' by Winston Churchill, *Tirpitz* was a 52,000 ton monster, armed with 15 inch guns capable of 22.4 miles' range, while it could go as fast as 34 knots, posing a major threat to Allied shipping in the Atlantic, as well as Artic convoys to the northern Soviet Union. Churchill declared her destruction 'of utmost importance'. By early 1944, she had already seen off a succession of raids, involving almost 400 bombers, torpedo-bombers, fighters and reconnaissance aircraft. The last attempt almost succeeded, when mini submarines laid charges on her, which when exploded lifted the ship 6 feet out of the water. Still, she did not sink. Although she was badly damaged, intelligence indicated she would be back in action by spring 1944, by which time it was hoped the Mosquito would be ready and waiting. If this was a success then the Americans also wanted to utilize the Mosquito to bomb the huge new battleships of the Imperial Japanese Navy, such as the 65,000 ton *Yamato*.

Despite the importance of the operation, and the tight turnaround, this wasn't Eric's only duty. He still had other projects to work on, including testing a second captured Focke-Wulf Fw 190A-4 and the Hawker Typhoon Mk Ib, which was needed for the forthcoming invasion of Europe. In the Typhoon, Eric was nearly 'kidnapped' by a fake German radio homing transmission, which took him across the Channel and over Occupied France. Eric suddenly became aware of the ruse when he saw the French coast though a gap in the clouds. It was a close shave and it would have delivered a new aircraft type, and an experienced test pilot, directly into the hands of the Luftwaffe.

With so much work, and all against the clock, even Eric admitted that it put an 'extra strain' on him. Such pressures weren't for everyone. For the trials he was initially joined by Brian Bullivant, another Fleet Air Arm pilot, but due to the dangers and stresses of the job he only lasted a month. Despite this heavy workload, the special trials for the Mosquito deck landing were given priority, and Eric wasted no time in starting preparatory work to understand the aeroplane and its flying characteristics.

Eric's job was made somewhat easier, as de Havilland at Hatfield and Marshall of Cambridge had modified the standard Mosquito bomber into a special naval version, which had a Barracuda-style hook, and a strengthened after-fuselage, as the original aeroplane had not been designed to take the shock of a deck landing, nor the sudden stop of the arrester wires. Eric was subsequently assigned to fly serial number LR359, with flight test engineer Bill Stewart in the right-hand seat, instead of the customary navigator of a front-line Mosquito.

Stewart and Eric had been introduced a few days earlier, and their common Scottish heritage created a useful working relationship, which continued until Eric left the RAE in 1949. 'Bill was from Glasgow, and there were times when we would tease each other,' Eric recalled. 'Being Edinburgh educated and having played rugby against Glasgow University, I would remind him of whom was usually top!'

Before the trials began, Eric and Stewart had plenty of literature to get through, and it did not always make happy reading. 'The problem was that the pilot's notes for the Mosquito gave the landing speed of the Mk VI as 125 mph,' Eric said, 'but the arrester wires' limit on the new Fleet Carriers was just 83 mph and the Mosquito stalls at 110 mph.' And there was another issue. 'We had more powerful Merlin engines and new four-bladed propellers to give both better clearance on the flight deck and allowing more power at a lower speed,' he said. This was all well and good, but 'the new propellers could not be feathered in the event of an engine failure, so there would be drag (which would impede the forward movement of the Mosquito).'

Perhaps the most frightening issue of all was that, in the event of losing an engine, the Mosquito had a tendency to flip over on to its back. This would be disastrous on a carrier and surely fatal. Eric knew that, to avoid this, the critical phase would be the last stage of the approach to the flight deck.

After some successful dummy deck landings, including stalling the Mosquito with both engines functioning and, then at 160 mph, on only one engine, having throttled back the propeller of the other, more intensive trials were scheduled on 18 March, at Crail and East Haven. This boasted an innovative dummy deck, with a mobile 'island' superstructure, which could be used to simulate the confined nature of an aircraft carrier flight deck. Space was very tight, with often only a few feet to spare between the landing aircraft's starboard wing and the superstructure.

Bob Everett had been selected as the deck landing control officer, or 'batsman', for the Mosquito trials and relished the job, much to the initial disgust of Eric, who really didn't think he needed a batsman at all. For once, Eric quickly realized he was wrong. If the Mosquito was going to make it on to the flight deck, someone would have to give the 'cut' signal, and Everett was the best deck landing control officer in the Fleet Air Arm. 'He was a natural showman,' remembered Fleet Air Arm veteran Ralph Jameson, 'when he walked into the wardroom, any wardroom, everyone knew he was there. He loved amateur dramatics and show business.' However, with the *Tirpitz* soon set to be back in action, and dummy deck trials complete, it was now time for the real thing.

With time running out, on 25 March, Eric was set to land the Mosquito on HMS *Indefatigable*. As he said himself, it was his 'first big test as an experimental pilot'. For this there was 'absolutely no precedent'. Eric certainly had no concerns about his flying ability, but he still remained unconvinced that the adapted Mosquito would hold up. The impact on a metal machine would be bad enough, but with it being so heavy, and made of wood, the consequences didn't bear thinking about. To make matters worse, a small army of boffins, top brass and VIPs would all be aboard HMS *Indefatigable*, watching on. If things went wrong, then his reputation, one he had taken so much care to build, would be in tatters. To Eric, this was his opportunity to prove that he, Eric Brown, an orphan from Edinburgh, belonged in such illustrious company, and it meant everything to him.

After lunch at Machrihanish, Eric boarded his Mosquito and set off for HMS *Indefatigable*, which he soon spotted between the bottom tip of Arran and Ailsa Craig. With the wind over the deck registering at 46 mph, Eric brought the Mosquito around for a circuit. Settling into the pattern at 600 feet – having already established that

500 feet would be the minimum go-around height, and 170 mph would be the minimum speed for the downwind leg, Eric saw Bob Everett give his signal to land. It was now or never. Reducing the speed to 82 mph, he crossed the flight decks after end, known as the 'round-down', and took No. 2 wire at 78 mph, before coming quietly to a stop. Eric switched off the accelerometer, and then the engine, before taking a moment to check himself. He looked across at Stewart and smiled: the landing had been perfect, and the Mosquito had held up. He had done it.

Upon climbing out of the Mosquito, Eric was instantly met by a swarm of boffins and those from the higher echelons of the Admiralty. All braving the wind on deck, they happily patted Eric on the back and shook his hand. It had gone better than Eric had dared to dream. This was his moment, and he was determined to cherish it. But there was little time to celebrate.

After a coffee break, during which the engineers checked over the Mosquito, it was soon time to make history once more, this time testing the Mosquito's take-off capabilities from a carrier. With the Mosquito pulled back to the after end of the flight deck, to give it the full length in which to take off, Eric admitted that he was more concerned about this than the landing. As the Mosquito's wingspan was so large, and in danger of hitting the island, he lined up to the left of the flight deck centre line. This meant he would miss the island, but there was also the prospect that the engine swing might cause the Mosquito to fall off the deck and into the ocean. Eric accepted the risk. This was the only way it could be done.

Putting on the starboard rudder, to counter the swing, Eric opened the throttles and charged down the runway. Moments later, he was in the air, climbing high above the ocean and the carrier's island. He had just carried out the first twin-engined take-off of a British aircraft from a British aircraft carrier.

With more test flights scheduled the next day, Eric was due to spend the night on board HMS *Indefatigable*, where a small party was held to celebrate the day's historic events. However, if Eric expected this to be the night he would finally feel as if he truly belonged, then he would be sorely disappointed. While all the boffins and VIPs were assigned cabins, Eric found that there was not one for him. The man who had risked his life, and made history that day, had to make do with the wardroom couch. It was a slight that Eric would not forget and left a bitter taste.

Woken early by the cleaner and ordered out of the wardroom, Eric was hurt and angry when he had to prepare for the next day's tests. With Bill Stewart flying alongside Eric as a supervisor, he was scheduled to make three more landings at steadily increased weights. The first two landings at 16,000 lbs and 17,000 lbs went well, but the third, at 18,000 lbs, finally proved too much. On landing, the arrester hook sheared, and as the aeroplane flew down the deck towards the island, Eric had just a split second to open the throttles and 'almost dived over the carrier's side' as he flew back into the air. 'People say that my wheels touched the surface of the sea, but I think not,' Eric said with a characteristic chuckle, if not overwhelming relief, as he took the Mosquito into the sky and away from danger.

After losing its tail hook, the damaged Mosquito was taken off the project, and a spare aircraft brought up from Cambridge.[1] The second Mosquito was also converted with an arrester hook for subsequent trials. In fact, it had been standing by at Cambridge should the initial trials fail – that is, should the first converted Mosquito crash into the sea – and another test pilot, Lieutenant Powell, was ready to fly it north to carry on. This second Mosquito was also written off after trials when it crashed on landing.

As it happened, the Mosquito wasn't needed to sink

1 Marshall of Cambridge was contracted to carry out the actual work, as the company had developed a good working relationship with both de Havilland at Hatfield and the Royal Aircraft Establishment. At Hatfield, the de Havilland factory concentrated on production and companies like Marshall developed one-off prototype modifications under the de Havilland engineering umbrella.

Tirpitz. After a succession of efforts, on 12 November, twenty-nine Lancaster bombers were finally able to destroy the ship just off the coast of Tromsø, taking 1,000 German sailors with her. The war also finished before two aircraft carriers were ready to be sent 'east', with twenty-four specially modified Mosquitos on board, to find and destroy the Japanese warships. This might have been fortunate, as landing accidents had plagued the Royal Air Force bomber crews of No. 618 Squadron, earmarked for these attacks. When they practised aerodrome dummy deck landings with Fairey Barracudas, five were so badly damaged that they were written off. 'I considered the Barracuda to be the easiest aeroplane to deck land, but they just couldn't get the hang of it,' commented Eric. If they had found the Barracuda such a challenge, then landing the Mosquito might well have proved impossible to master for the average pilot.

However, eighteen months after the successful Mosquito tests, Eric was asked to fly and test a fast and agile single-seat version of the Mosquito called the Sea Hornet. He was impressed. 'All the faults that I found flying the Mosquito for flight deck operations had been rectified,' he recalled, 'and I found the Sea Hornet the most perfect of fighters to fly and deck land.' Thanks to Eric's tests, it was decided to build sixty-seven 'Sea Mosquitos', although these barely saw any action and were retired from service in 1953.

The Mosquito tests had proved a high point in Eric's career to date, with him being awarded the Air Force Cross for his efforts. And yet his treatment aboard HMS *Indefatigable* seemed to confirm that, no matter what he did, he would always be an outsider. Typically, rather than shy away, Eric only became more determined than ever to prove his worth. It would come at a considerable cost.

Hitler's terror weapons had a profound impact on Eric's life. At work, he was assigned to testing new fuels, tactics and aircraft types. At home, in June 1944, his married quarters in Farnborough were hit by a doodlebug, almost robbing Eric of the love of his life. The house was destroyed, the dog ran away and the cleaning lady was badly injured. Luckily the gin survived.

15

Doodlebugs

In the early hours of 13 June 1944, the skies of London rumbled to the sound of several small aircraft rocketing above the East End before plummeting to the ground in Bethnal Green. An enormous explosion ensued, destroying buildings and killing six civilians. The raid was initially attributed to Luftwaffe aircraft, but this was far more serious. Britain was really being attacked by pilotless flying bombs.

Since the 1930s, the Germans had been working on flying bombs and rockets, but it was only in 1943 that developments moved apace. With German cities under constant attack, Hitler needed something to strike back at Britain with, and this is where the so-called Vengeance Weapon 1 (V1) came into play. Made of little more than sheet metal, and loaded with explosives, the V1 was propelled by a simple pulse jet engine that ran on 80 octane gasoline. Fired from German positions in France, or air-launched from Heinkel bombers over the Channel, its 148 mile range meant London could be easily targeted, without the need to risk pilots. This looked to be a new frontier in the Battle of Britain, which was long thought to be over.

Initially, British and American bombers successfully targeted the French launch sites, but the Germans were able to replace and camouflage them, quickly making them operational once more. Until the Allies in Normandy could overrun the launch sites, Britain could only rely on a layered defence of fighters, anti-aircraft guns and barrage balloons close to the capital. Yet these weren't always effective, as the flying bombs were as fast as 400 mph and unpredictable. Over the course of the next three months, the Germans launched more than 8,000 V1 strikes, nearly all of them against London, killing 5,500 people, injuring 16,000, and forcing the evacuation of more than a million people, with one flying bomb even narrowly missing Buckingham Palace. The distinctive sound of the V1's pulse-jet engine led to the British calling this terrifying new breed of weapon the 'doodlebug'.

At this time Eric's hands were already full. The pace of work at Farnborough was frenetic, and he would often fly up to ten different aircraft in a day, some just for a ten-minute circuit to prove some modification created by the boffins. He was also undertaking vitally important handling trials on the four-engined Handley Page Halifax bomber, helping to successfully test a new tail and fin after it had suffered issues in night-time operations over Germany. His co-pilot for these trials was Leonard Cheshire VC, a distinguished bomber pilot who would command No. 617 (the Dambusters) and who pioneered postwar rehabilitation centres for those coming to terms with injuries and disabilities. From his notes, Eric was in awe of Cheshire, and from his comments, Cheshire was in trepidation of a naval test pilot in a four-engined bomber. During this period, the work was often dangerous, none more so than during tests on a Seafire Mk IIc (a navalized Spitfire). Propelled by eighteen rocket motors on an elevated track, on take-off the trolley failed to separate correctly. This meant that Eric somehow had to land the Seafire, with an extra 100 lb weight of its catapult bridle wrapped around the tail-wheel launch-spigot. Geoffrey Cooper, who witnessed the event, recalled:

> Instead of flying a normal three-quarters circuit after the bone-shaking launch, assisted by un-rotating projectiles – rockets, in plain speak – Eric flew a complete circuit to land back on the main runway within the airfield boundary . . . Eric was a master in the air, he just sorted out the aircraft's trim to fly straight and level and returned to the runway, where the attachment dropped off as he bounced on landing.

For this brush with disaster Eric always blamed Norman Skelton,[1] the technician responsible for filling the braking system for the trolley with water. He felt that Skelton was overawed by the task because a party of VIPs were watching, and he did not add sufficient water to act as ballast. One of the bystanders who so overawed Skelton was Winston Churchill. It was another of Eric's close shaves in front of the PM.

However, while Eric continued to work on a number of different projects, one of his most vital was to test the ability of low-flying fighters to intercept and stop the doodlebugs. Farnborough was therefore tasked with testing all of the current fighters' capabilities for such an operation. Yet as Eric commenced testing, fate struck. He was informed that his new quarters at Manor Road had been hit by a doodlebug.

Immediately rushing home, Eric feared the worst. From a distance, he could make out a mound of rubble where his house had once stood, and where he knew Lynn had been that day. To his intense relief, as he drew closer, Eric saw his soot-covered wife standing on top of the smouldering remains of the Brown family home, singing and entertaining the Hampshire Salvage Corps crew. After the bomb had hit, Lynn was so shaken that

1 In some of his talks, Eric called Skelton 'Murphy' to exemplify 'Murphy's Law'.

she had 'finished a bottle of gin, which was about the only thing that she could salvage from the destroyed dining room'. By the time Eric reached her she was already 'very merry' but thankfully unhurt. Sadly, the bomb which detonated in the garden caused the 'daily help' Alice to lose an eye and the pet dog Winston to run away and hide for days.[2]

Back at Farnborough, Eric and the team became ever more determined to defeat this menace.

They realized that the key lay in somehow increasing aircraft speed to catch the bombs, which were too quick for all but the latest fighters. A super high-octane fuel was therefore developed to be used for the already high-performance Spitfire XIV, Mustang III and the new Tempest Mk V. This was, however, extremely dangerous. The fuel was volatile and in danger of exploding. As always, all eyes turned to Eric to see if it could work.

On 26 July 1944, Eric put the new Hawker Tempest Mk V to the test. This fighter was a direct successor to the Hurricane and Typhoon, marking the zenith of piston-engined fighters from Hawker at Weybridge. It was already the fastest piston-engined fighter of the war at low altitude, but the 150 octane fuel would ensure that it had sufficient speed advantage to catch and engage the V1. All

2 In another version, Eric wrote that the kennel was the point of impact, and the dog perished.

was going well until at 7,000 feet Eric smelled burning oil from the floor of the cockpit. He looked to the dials and saw oil pressure and oil temperature were both at zero. This wasn't good, but for now Eric could still fly, so before things got any worse, he decided to try and land. Yet as he came down to 4,000 feet the engine misfired and the propeller oversped. Moments later, a loud bang rocked Eric in his seat, as oil sprayed over the cockpit, which quickly went up in flames. Somehow, Eric kept his nerve. The test pilot's first instinct, rather than to bail out, is to get the stricken aircraft back to an airfield for the engineers to find the fault. But as he turned for Farnborough, Eric knew things were getting serious when the heat became so intense that he had to take his feet off the rubber pedals. Still, he refused to bail out until the propeller seized solid, and the belly of the fighter went up in flames. Only then did Eric admit defeat. He had to get out, and fast.

Down to between 1,600 and 1,200 feet, and travelling at 170 mph, he took off his seat harness and with a struggle managed to open the cockpit, with the ground rapidly coming into sight. Putting his leg over the port side of the aircraft, his face being buffeted by the wind, he steadied himself to jump when the Tempest banked sharply and threw him out. Tumbling towards the ground, Eric grasped frantically for his parachute rip cord, finally managing to pull it with less than 1,000 feet to go. It was his first parachute jump.

Sailing to the ground, Eric's ordeal was not yet over. He landed in a duck pond, and while it was thankfully shallow and allowed Eric to wade out at waist height, he then found a bull, pawing the ground, waiting to confront him in a farmer's field. Eric had no option but to try and remain calm as the emergency services, and the farmer, soon came to his rescue.

Thankfully, the boffins were able to overcome this fuel issue and future tests were far less dramatic. Soon after, Spitfires, Tempests and Typhoons, all loaded with 150 octane fuel, were able to intercept the doodlebugs. Once in range, they could either shoot down the V1 with a 20 mm cannon, ensuring they were at a safe distance to do so, or even slide a wing under the V1, and tip it over and out of control. It required careful handling but reduced the risk of flying through the debris of an exploding bomb.

By the end of August, not more than one bomb in seven got through to the London area. By October, the last of the launch sites within range of Britain was overrun by advancing Allied armies.

While Eric had been heavily involved in the ceaseless effort throughout the war to extract more speed from propeller-driven aircraft, a quantum leap forward for aviation was just around the corner. Once again, Eric would be at the forefront.

Ever since he saw the first flight of the British Gloster, on 13 May 1941, Eric was intimately involved in the development of jets. His experience with German, American and later British developments led to the world's first ever deck landing by a jet aeroplane on an aircraft carrier, on 3 December 1945. It was a remarkable achievement and ensured Eric another entry in the record books.

16

The Jet Age

During the early months of 1944, the Directorate of Naval Air Warfare in the Admiralty was urgently considering the future of naval aviation. By now aircraft carriers had established their place as the world's capital ships, the role previously held by the battleship for a century, but to keep pace it was realized that using jet-powered aircraft would have to be explored. Recent events in Germany had also made this a real priority.

In mid-1944, the German Messerschmitt Me 262 had come into operation. It was considered to be the most advanced aircraft of the Second World War. Thanks to its swept-back wings and two axial-flow jet engines, it was at least 100 mph faster than any other aircraft. The Allies needed a fighter to match it, and fast.

As such, by March 1944, a comprehensive jet test programme was underway at Farnborough, something which certainly enthused Eric. He was understandably excited about any great leap forward in aviation design, especially as in this case he also had a personal connection to it.

Just a few years earlier, on the afternoon of 13 May 1941, Eric had set off for Croydon Airport, where he was to have new shoulder harness straps fitted to his Martlet fighter. A series of landing accidents had caused facial injuries to pilots who were wearing the American-style lap-strap, and modifications were subsequently ordered fighter by fighter. Eric had been one of those injured pilots and he was delighted with the prospect of both the flight down the length of Britain to Croydon and the better safety equipment. However, due to poor weather, Eric was forced to stop at Cranwell in Lincolnshire.

As the weather worsened, Eric was resigned to spending the night there. At a loose end due to this unexpected turn of events, Eric and his roommate, Geoffrey Bone, an engineer from the Royal Aircraft Establishment, planned a trip to the cinema in nearby Sleaford. With fuel on military airfields not metered in those days, and the trip requiring filling Bone's car, it was felt a few gallons wouldn't be missed. However, the fuel was not the standard vehicle, or even piston-engine aircraft fuel, but a new type called AVTUR – aviation turbine fuel, more like paraffin than petrol. Not knowing this, Eric was mystified when the car back-fired, 'pinked' and generally ran more like a tractor than a motor car. The nature of the fuel, and the reason for it, only became evident the next day.

While eating breakfast in the officers' mess, Eric didn't notice renowned Gloster test pilot Flight Lieutenant

Gerry Sayer sitting across from him. Even if he had, he still would not have guessed why Sayer was there. Around midday, after flying a series of weather checks above the Cranwell airfield, Eric did at least notice a lot of commotion around a hangar. It not only had an RAF Police guard but also a large number of civilians swarming around it. He took the civilians to be from the 'Ministry', or even from one of the manufacturers, but no one would confirm – or deny – who they were or what was happening. This wasn't unusual, and in wartime you didn't ask too many questions.

After lunch, the weather forecast was still too bad for Eric to continue to Croydon. The Senior Air Traffic Control Officer took him to one side and instead suggested he go for a cup of tea, then return to the tower's glass-windowed control room, where he might just witness history taking place. He said no more but warned Eric to ensure he stayed out of sight. Eric had no idea what this was all about, but he was grateful for the short break and intrigued by the secrecy.

Eschewing early supper, Eric waited quietly in the flight planning office, where, in any case, he needed to plan for his departure south the following morning. By 1900, the sun had come out, and Eric noticed that the hangar doors, previously roped off, had opened to reveal a curious, small aeroplane without any sign of a propeller. He was bemused. He had seen nothing like it before. By sheer luck, on that day, 15 May 1941, he just happened to be present at RAF Cranwell to witness the inaugural flight of Britain's first jet-powered aeroplane, the Gloster E28/39. It was for this reason that AVTUR – aviation turbine fuel – had been on base.

Although Eric couldn't clearly see the people gathered at the hangar that day, the gaggle around the new aeroplane included not only test pilot Sayer and engineer Bone, but the inventor of the jet engine, Group Captain Frank Whittle. For Whittle, this was the culmination of over a decade of work.

In January 1930, Whittle had finally realized the potential of his gas turbine patent, which was subsequently tested at Lutterworth on 12 April 1937. The jet age was born that day, and Whittle was the father. The Air Ministry specification for the so-called Gloster E28/39 (later given the less cumbersome name of 'Pioneer') was issued in 1939, prior to the declaration of war on Germany in September. The specification included reference to its intended secondary role as a day fighter:

> The primary object of this aeroplane will be to flight test the engine installation, but the design shall be based on requirements for a fixed gun interceptor fighter as far as the limitations of size and weight imposed by the power unit permit. The armament equipment called for in this specification will not

be required for initial trials, but the contractor will be required to make provision in the design for the weight and space occupied by these items.

Manufacturing of the airframe began at Brockworth, near Gloucester, but was moved in secret to a former motor car garage at Regent Motors, Cheltenham, to avoid the attentions of the Luftwaffe. Gloster's George Carter, and his team of engineers and designers, worked closely with Whittle, ensuring the first prototype was completed a few weeks before the maiden flight in 1941.

Although E28/39 had made several test hops at the Gloster airfield at Brockworth in April, it was only in May that a flightworthy jet engine was available. The decision was made to take the prototype to Cranwell, which was considered a secure site, where there was plenty of take-off space in any wind direction. Cranwell was also in sparsely populated Lincolnshire, and although the county would eventually see twenty-seven airfields in wartime use, it was still off the beaten track and away from prying eyes. Coincidentally, Whittle had been a cadet at Cranwell when he had first thought about the jet engine and the propeller-less aeroplane.

That first flight, which Eric witnessed, lasted just seventeen minutes. The roar of the engine as it thundered across the grass was unlike anything Eric had ever heard, while its seemingly effortless progress through the air clearly pointed to the future. He was stunned that an aircraft with no propeller could move so fast. It looked like something from outer space.

Within the same month, after several more flights and some engine modifications, the Gloster jet reached 25,000 feet and was well on the way to being handed over to service pilots at Farnborough for further development. Before test flying was concluded, the E28/39 would fly as high as 42,000 feet, and at over 500 mph. And it did not go unnoticed by the Admiralty.

By May 1944, Farnborough was tasked with testing how jet-powered aircraft might be utilized on aircraft carriers. There were lots of things to consider, particularly the structural integrity of the airframe on launch and landing. At sea, too, there were the dangers of bird strike to a pilot sitting in the nose without the protection of a piston engine in front of him. The risk of water ingestion was also a concern, as the sea could wash over the bow and into the engines of aircraft parked on the flight deck. Would the water stop the jet engine functioning at a critical time in flight?

With Eric ordered to take the first jet to sea as soon as possible, the design would be crucial. 'We were on a steep learning curve, and every hour of flight test counted,' he said. Thankfully, by this stage there was a handful of different designs Eric could consider at Farnborough. These included the Gloster E28/39, as well as other prototypes, including the new Gloster Meteor, the American Bell Airacomet and a new design from de Havilland.

Again, the dangers to Eric soon became clear. During flight testing of the new jets their engines often failed, leading to Eric trying to land at Farnborough or more often than not put down wherever was safe. He couldn't deny, however, that flying higher or faster than ever before was a real thrill.

After some serious test flying, Eric needed to reach a decision. Following discussions with Farnborough and plane-makers' staffs, it was decided that there was little benefit from modifying and then taking the experimental Gloster E28/39 to sea. The Gloster Meteor was also, for the moment, lacking in power and required too much runway for take-off (a luxury not available at sea). Eric described the Mk I as 'an underpowered aeroplane with heavy controls [which], in my opinion, would have been easy prey to the Me 262.' The Bell Airacomet was also not suitable for the task, described by Eric as 'a terribly ponderous aeroplane'.

Running out of options, on 19 May 1944, Eric flew a prototype of the de Havilland Spidercrab, later to be known as the Vampire. The engine took time to respond to throttle inputs, but he felt its docile handling would work well on a carrier. In Eric's words, 'She performed excellently.'

The de Havilland team, led by Major Frank Halford, had experience in simple, lightweight airframes, of which the wooden Mosquito was the outstanding example. They had also conceived a different engine design from that of Whittle's Power Jet Company, which was used in the E28/39 and the two-engined Meteor. Halford's 3,000 lb static thrust engine was ready in 1942 and as a result, the Air Ministry issued Specification E6/41 for two prototypes of the DH 100, aka the Vampire.

A little less than a year later, Eric began dummy deck assessment of the Vampire at Farnborough. The dozen or so runs showed that it could be handled well on the dummy deck, even if it tended to land slightly off centre-line. In a rare moment of self-doubt, Eric admitted that this off-centre line landing might have been his fault.

However, by the summer of 1945, with more trials scheduled before the Vampire was ready to be tested on a carrier, the war was over. Nevertheless, while Eric had many postwar duties to attend to, which we shall come to in due course, he continued with the trials.

On 11 October, arrester hook trials started, only after considerable work was needed to define the exact position of the arrester hook on an aeroplane with a twin tail boom and short fuselage configuration. The hook was eventually fitted near the end of the fuselage.

On 17 October, there then began a series of forty aerodrome dummy deck landings on a mock-up deck to represent that of the brand-new aircraft carrier HMS *Ocean*. Creating the landing routine, and providing Eric with a pilot's eye view from the ground, was the experienced Deck Landing Control Officer Lieutenant-Commander Jimmy Pratt. There was only one mishap, when the port wing touched the ground on the round-out of the landing approach; this was probably due to the Vampire's slow speed and a gust of autumn air.

By early December, the land trials were complete, and at last the little Vampire was ready for sea, with HMS *Ocean* prepared and standing by. The carrier was commanded by veteran naval pilot, Caspar John, who just happened to be an Eric Brown supporter, a relationship which would prove crucial in years to come.

On 3 December, operating from the STU hangar at RNAS Ford, Eric prepared to fly the Vampire out to HMS *Ocean* in the Portsmouth Exercise Area. He carried an envelope that Lynn had handed to him when he left that morning, inscribed; 'To be read immediately before take-off'. Tearing open the envelope, Eric read the letter:

My dearest darling!

I am writing this while you are sitting there unconcernedly lighting your pipe, completely unaware of what I am doing.

This is just a hasty note to wish you all the luck & success in the world. I do so wish I could be there to see you do your aerobatics, then the last circuit, and then (greatest thrill of all!) to watch you come floating down as light as a feather, to touch down with lots of deck to spare, while all the troops are cheering, & the 'gold' are swelling with envy.

This is your Big Adventure, sweet; your little bit of history is in the making. I am so proud of you & all day on both days I shall be thinking of you, & keeping my fingers very tightly crossed.

Look after yourself, sweet – you are precious to me & the country & neither of us can spare you. Hurry, hurry back as I shall be waiting for you for our own special kind of celebration.

Love, luck, & all power to you.
Your loving Lynn.

This letter, the first and only one of its kind, showed just how worried Lynn was for Eric at this time. He had faced danger before, but even she appreciated just how much this would test her husband to his limits. From her writing it is clear that the two must have discussed all of these concerns beforehand, while it also shows the support which Lynn gave to Eric during his test flying. Whilst Eric was able to compartmentalize his fears and worries, devoting himself to test flying and disregarding all else, Lynn was often left at home to ponder. She was a constant support, even if the worry at times must have been overwhelming. But with test flying in this new jet age, Eric was certain this would finally see him make his name and gain the recognition he craved. He aimed to grasp the opportunity with both hands, even if the risks were high.

Yet before Eric could make history, there was an issue. The weather that morning was grey and dull, so much so that the Admiralty decided to postpone the trial and sent a signal to that effect to the operations staff at Ford. Whether it reached Eric before take-off or whether he chose to ignore it is unknown. What we do know is that he had a very good reason for wanting to attempt the first jet landing on a carrier as soon as possible. 'I was desperately keen to beat the Americans, at being the first to operate jets from carriers,' he admitted in his autobiography. 'I did not want to risk delay by damaging the machine on the airfield. I just could not wait.'

Whether he wilfully ignored orders it is hard to say, but in any event, Eric took off in the Vampire and made his way to HMS *Ocean.* At about 11:15, Eric spotted the carrier below. However, flying in the wind was a flag from the gaff on the island superstructure, telling all aircraft to stay away. Still, Eric would not be deterred from claiming his place in history. So, not for the first time, he ignored orders and decided to land, even though the carrier was not at flying stations.

Thankfully, as the Vampire rocketed around HMS

Ocean, Captain Caspar John understood that Eric had no intention of waiting around. When he saw the diminutive jet pass down the ship's side, he ordered the carrier to turn into the wind; the Channel breeze and ship's speed created 42 knots over the deck, slightly more than the Vampire's stalling speed with full flap deployed. It was enough. Settling into his final approach, Eric realized that the carrier was moving more violently than he had expected. Fighting to keep the Vampire steady, he hit the deck at 11:28, picking up No. 1 wire, pulling it out 106 feet, while decelerating at a stately 1.05 g before coming to a stop. The world's first jet landing on an aircraft carrier was complete, and Eric had flown into the history books, again. 'We had succeeded,' Eric triumphed. 'We had arrived.'

Waiting to greet him on the flight deck were senior naval aviation officers, boffins and officials who had raced up from the wardroom to watch when the Vampire had unexpectedly arrived. The first down on to the flight deck to shake Eric's hand was Vice-Admiral Sir Denis Boyd, the Fifth Sea Lord and head of naval aviation. He was then congratulated by Arthur Woodburn MP, the Parliamentary Secretary at the Ministry of Aircraft Production, de Havilland's chief designer R. E. Bishop and the assistant chief engineer, Richard Clarkson.

In the face of history, everyone swiftly forgot that Eric had landed the jet against orders to postpone the test. Indeed, Admiral Boyd later sent Eric a personal signal to congratulate him: 'Warmest congratulations on your skilful handling of the Vampire and the successful achievement of such a notable event in the history of naval aviation.' HM King George VI also recognized the feat of airmanship on 19 February 1946 by investing Eric with the OBE, military division. In the event, the King was unwell but wrote Eric a personal letter of regret for not being there in person.

A further eleven landings followed at sea over two days, with the trials regarded as a roaring success. Eric was now fully in the sights of the aeronautical community and on 10 July 1947 was invited to become a full member of the Naval Aircraft Research Committee, based at the National Physical Laboratory at Teddington. Unbelievably, Eric was still just a Temporary Lieutenant-Commander (A) in the Royal Naval Volunteer Reserve, yet he was now in the inner circle, or so he thought.

However, despite this success, some problems remained with operating a jet at sea. The previous piston-engine-powered fighters had been able to 'hang on the prop' and effectively stall on to the flight deck to catch the arrester wires. In contrast, the jet pilot had to maintain power to avoid the stall until the arrester wire had been caught and the jet's progress halted. If not, the barrier beckoned, and the steel cables of the conventional barrier, strung across the flight deck, would be lethal for the pilot in the nose of the speeding jet. Furthermore, for all practical purposes, the forward deck park of the flight deck was unusable for other aircraft due to the risks of a jet going through or over the barrier. This created parking problems and made fluid flying operations almost impossible.

At Farnborough, Lewis Boddington and Eric worked on possible solutions. After much deliberation, Boddington came up with the idea of having separate decks for launch and recovery: a flying-off deck on the hangar deck (below the flight deck) with an open bow, similar to the design of HMS *Furious* in the 1920s. Landing would then be on the flight deck, on a rubber 'mat'. This radical solution meant that an aircraft with a heavy undercarriage was not required.

The rubber mat was the idea of Major F. M. Green, the former chief of design at RAE. Of course, it became known as the 'rubber deck' and was a large undertaking. 'It covered the width of the flight deck from aft of the island [superstructure] and consisted of a huge sheet of rubberized multi-fabric ply, very like a motor tyre [in] construction,' wrote Captain Colin Robinson, then a Lieutenant ME (mechanical engineer) aboard the trials ship HMS *Warrior*.

Before it went to sea, there were trials scheduled throughout December 1947 with a static version of the rubber deck at Farnborough. There was, however, a frustrating wait due to bad weather, which seemed to go on for weeks. When the day for the trial finally came, it would be one of the worst days of Eric's life, for more reasons than one.

On the morning of 29 December, Eric's father, who had been staying with him over Christmas, was taken ill. While Eric waited anxiously for an update from the doctors, a call came through from Charles Crowfoot, who ran the flight tests in the Naval Aircraft Department. The conditions were finally perfect for the first attempt to land on a rubber deck. At this time Eric's father was still conscious and knew that his son had been anxious to test the rubber deck for the last few weeks. He told Eric to go ahead and he would see him when he was back, fresh from making history.

Just after midday, Eric performed one dummy run, then brought himself in to land. Perhaps Eric was distracted by the morning's events but he approached the mat too quickly and landed heavily, bouncing over the rubber deck and damaging the Sea Vampire trials aircraft. Eric was lucky not to be seriously hurt as the impact split the cockpit.

After a quick stop at the sickbay, where he was checked over and declared fit, Eric returned home, only to find that his father had just passed away. It was a devastating blow. Whatever the truth behind Robert's Royal

Air Force fantasies, he was still Eric's mentor. He had given him a home and a loving family, encouraged his love of aviation, then introduced him to the pre-war excitement of Germany. Eric never treated Robert as anything other than his real father, but with his passing he suddenly felt very vulnerable and alone. His adopted mother and father were now both gone. Once more, he was an orphan.

As always, Eric dealt with such upset by throwing himself into work, particularly in trying to crack the rubber deck trials. In response, changes were made to both the aircraft's flying profile as it approached the deck and the construction of the rubber deck itself. Thanks to these adjustments, on 17 March 1948 Eric once more attempted landing the Vampire on the rubber deck at Farnborough. This time the results were far more successful. Such was the perfect landing that Eric described it as 'like a body hitting a tightly stretched trampoline'.

The next day, William Perring, the director of the Royal Aircraft Establishment, sent Eric a note saying, 'I hear it was a perfect landing. Everyone will now be anxious to see you do it again.' No pressure, then. Eric thought enough of the note to paste it in his photograph album and fulfilled the director's wish with fifty-two successful landings – only then was the design ready to go to sea.

On 3 November 1948, the first rubber deck sea landing was set for the light fleet carrier *Warrior*, where the rubber deck had been fitted 2 feet 3 inches above the steel flight deck. After two dummy runs, Eric brought the Vampire in at 118 mph. The landing was a little uncomfortable, but there was no damage to the aircraft, or Eric.

From mid-1948, trials went on almost every weekday for a year, and the ship's company grew used to the nail-biting few seconds of the Sea Vampire disappearing over the bow before the pilot gained flying speed and reappeared. Eric commented that this was all part of the 'fun of naval flying'.

However, after nine years of design and trials, even though the US Marine Corps showed interest for jungle operations, the rubber deck was abandoned as impractical. Eric was disappointed: he felt that the rubber deck had many benefits and would have considerably reduced jet landing accidents.

As well as the rubber deck not coming to fruition, it also took until 1950 for the Admiralty to put jets to sea. This decision was controversial and has caused much debate, with Eric often in the firing line. Veteran naval aviator and former STU officer Commander David (Shorty) Hamilton certainly blames Eric for the delay. In a research paper, he quotes Eric's report, which was circulated to the STU, and said:

> His report concluded with the following . . . 'it was thought that operating the aircraft with the combination of no stall warning and engine handling problems would tax pilot skill' and 'but it is hardly the final answer since the unreliable human factor is being taxed to the utmost in making up what the aircraft lacks – small but very important deficiencies'.
>
> By implying that the average deck-landing pilot would have insufficient skills his statement put back the introduction of the jet for the (Royal) Navy . . . he knew that no Fleet Air Arm aircraft had stall warning [which] if anything could be very dangerous in distracting the pilot from the main task. He did not specify [what were] the small and important deficiencies.

Hamilton is backed up by Commander Stan Orr DSC AFC, a fellow jet test pilot, based at Boscombe Down, who wrote:

> Eric had concluded that the exercise would be too difficult for the average pilot with the aircraft in its present state and therefore design developments and modifications, such as lift spoilers, would have to be built into future aircraft . . . [However,] the consensus of opinion [of three STU test pilots, including Orr] was that the deck landing . . . was simple and straightforward.

Hamilton maintained that Eric deliberately stalled the Vampire on to the flight deck, as he had dive brakes out and flaps extended, which, in the opinion of Hamilton and others, would make the Vampire more difficult to deck land than the straightforward approach. He also said in his note that 'extending the dive brakes and deliberately stalling the Vampire would, in fact, make a deck landing more difficult, especially if one had missed all the wires and had to open the throttle and go around again [for another landing].' Apparently, Eric managed to 'dink' the port wing flap on his fourth approach, which was 'not helped by stalling the Vampire.' 'For a pilot of his reputation,' Hamilton says of Eric, 'and in conditions of no hurry, no other aircraft, an aircraft carrier all to himself, it is surprising to say the least.'

Some of this might be attributed to professional jealousy, as Hamilton was a rival of Eric. However, Hamilton was not the only rival who expressed concerns.

Raymond Lygo was not only a rival but also had considerable jet experience to back up his opinions. During a twelve-month exchange with the US Navy, he had flown the McDonnell Banshee and had also embarked in the aircraft carriers USS *Coral Sea* and USS *Philippine Sea*. This was operational flying and not test flying, giving Lygo the accolade of being the Royal Navy's most experienced jet aviator by 1950. His criticism of Eric is based on experience but rather disingenuously fails to mention that the deck landing to which he refers was the world's first:

> The Royal Navy had yet to come to terms with jet aircraft; some years previously, 'Winkle' Brown had carried out some deck landings in a Vampire, after which he had, I believe, expressed the view that, with the engine control then available in a Vampire (which had a very slow reaction to throttle movement), the task would not be in the competence of the average squadron pilot.

This was test pilot versus operational pilot, and in Lygo's autobiography, *Collision Course*, he aims a harpoon straight at Eric: 'All human endeavour is sometimes set back by people who imagine they are so superior that none could match them and certainly never surpass them . . . within three years average pilots were doing it by day and by night.'

Lygo's comments led to two decades of friction with Eric, and, as we shall see later, their faltering relationship eventually came at a considerable cost to Eric's career. The controversy surrounding Eric and his achievements would only grow more divisive over time, particularly when he became more vocal and opinionated. Rather than listening to any criticisms and calmly debating them, he saw them as a personal slight born of jealousy that he, Eric Brown, had beaten them to the punch. In this case, Eric might have had a valid reason for the delay. 'The Vampire's fuel capacity was too small for carrier work,' he explained, 'and its radius of action would have been too small for carrier operations.'

Early jets were fuel-thirsty, and fuel-light, as the Americans found at the beginning of the Korean War. The Banshee and Panther had three times the fuel capacity of the Vampire and still struggled. In the end, the US Navy temporarily reverted back to piston-engined aircraft, such as the Douglas A-1 Skyraider.

Eventually, jet aircraft would, of course, transform military aviation. The German Luftwaffe had already proven that the speed edge which they provided would render piston-engined aircraft obsolete in many roles, particularly interception and other fighter tasks. For the Royal Navy, the adoption of jet aircraft was necessary in order to hold their own against land-based fighters and keep credible strike capability, which was the carrier's *raison d'être*.

For Eric's part in these trials, he was awarded the King's Commendation for Valuable Service in the Air in 1949. He also received the Fleet Air Arm's prestigious Boyd Trophy for 1948, in recognition of his rubber deck work. In addition, the Royal Aeronautical Society awarded him its silver medal for practical achievement in advancing the science of deck landing.

However, Eric's early experience of flying jets was not limited to new British designs, or indeed American ones. In the summer of 1945, with the end of the war fast approaching, a race was under way to capture and examine the Luftwaffe's latest aerospace technology. What they would find was extraordinary, and Eric was at the forefront of that discovery, which would also have a profound impact on the jet age, and beyond.

Winston Churchill was determined that Britain would come out of the Second World War well placed to exploit the advances in aeronautical technology developed at home and that created by the Germans. He ordered the setting up of a secret mission to Germany under William Farren, the director at RAE Farnborough. Naturally, he turned to Eric to lead the test flying, and Eric set his sights on the Messerschmitt Me 262 jet fighter, the fastest machine in service.

17

The Farren Mission

One grey day in March 1945, the Farnborough loudspeaker intoned: 'Lieutenant-Commander Brown is requested to report to the director immediately.' The director in question was William Farren, one of the nation's greatest aeronautical engineers, who had been an air-gunner in the Royal Flying Corps during the First World War and immersed in aviation ever since.

When Farren was appointed the first director of the Royal Aircraft Establishment in 1941, his reputation had gone before him. He had made his aeronautical engineering reputation on the development of wind tunnels, now so important for the advancement of high-speed flight, especially with jets, and where Britain had little pre-war expertise. He was also involved in highly classified work on the Whittle jet engine. Once he was in charge, things changed at Farnborough, and the whole 55 acre site started to buzz with new developments.

Upon hearing the tannoy, Eric walked across the concrete apron to the director's office, situated on the ground floor of the three-storey structure, designated

Building Q. It was a spacious room, with a good view of the airfield, and the office was so big that in later years it actually became the establishment's main library.

Eric was not worried about being summoned. He was on good terms with Farren, but still he couldn't help wondering what the summons to the Captain's table (as a naval officer might say) was all about. He quickly ran over the previous week's flying in his mind and couldn't think of any 'blacks', the misdemeanours that had occasionally blighted his early flying. He was older and wiser now: twenty-five years old, getting on a bit for a fighter pilot, but coming into his prime as a test pilot. So what was this all about?

'Come in, Brown,' Farren said, as Eric reached his office. He wore a serious but not unpleasant expression. 'Take a seat . . . tea?' Good, thought Eric, it is not a standing interview without coffee (as a naval officer might say). Still, something must be up, and indeed it was.

Farren went on to outline a highly classified brief. It centred on the Prime Minister's wish that, with the end of the war fast approaching, a mission was to be sent to Germany to 'survey German aircraft, engines and armaments, including the research and development organization and the route to manufacture'.

Churchill had set three main objectives for the mission:

1. Find, capture and interrogate German aircraft designers, senior scientists and test pilots.
2. Find, and prepare for flight test, as many new and interesting types of German aircraft as possible (Eric had identified fifty-three of significance).
3. Find and, if possible, dismantle any supersonic wind tunnels and transfer the technology and the physical parts to Britain.

Wind tunnels – large tubes with air blowing through them, used by engineers to replicate the actions of an object flying through the air – were deemed to be especially important. British designers had thus far neglected the benefits of wind tunnels, and Churchill was therefore determined to retrieve, lock, stock and barrel, the research wind tunnels dotted around the Third Reich.

Allied technical intelligence reports had also pointed to German progress in flying faster than the speed of sound, which Hitler thought would be a war-winning technology. While the Germans had been at the forefront of jet developments, since the mid-1930s they had also been working on tailless aeroplane designs, which they thought stood the best chance of breaking the sound barrier. Obtaining this aircraft, its designs and any other information to go with it was deemed vital if Britain was to be a world-beating commercial aviation nation in the postwar world.

However, with Germany's capitulation in May, the Farren Mission had to move fast. In the postwar era, allies were now competitors in obtaining Germany's advanced technology, particularly in the aviation field. Not wanting the British, or worse, the Russians, to gain key people and technology, the United States had launched Operation Paperclip. This American equivalent of the Farren Mission, with the highly effective Colonel Harold Watson of US Army Air Force in charge, was both a competitor and a collaborator and became known in America as Watson's Whizzers.

The mission was therefore a race to gather up the best and brightest people from the German institutions. This was not just important so as to gain a technical advantage for future commercial aircraft operations, but also to prevent the Soviets from developing new and inventive weapon systems, which might once more threaten the peace of Europe and the world.

In the face of such competition, the Prime Minister authorized Farren to use whatever means to beat his rivals and take with him to Germany whomsoever he thought would be needed. When the team was eventually assembled, it would involve many of the most prominent names in the British aeronautical industry. These included Basil Henderson of Parnall Aircraft; Vickers armaments guru Captain A. J. Nannini; Arthur Sheffield, who managed the English Electric works at

Preston; and the Scottish academic Dr Alec Cairncross, who would represent the Ministry of Aircraft Production and deal with global economic issues. Cairncross would also later play a prominent role in the rebuilding of Germany, living in Berlin for several months. From Farnborough, Farren chose Eric's immediate boss Morien Morgan, and George Edwards, another Vickers man, just promoted to the role of chief designer but with a wealth of experimental aircraft research knowledge. Naturally, Farren turned to Eric for one of the most important missions of all.

Farren directed Eric to create, and lead, the Farnborough Enemy Aircraft Flight and develop a methodology for the retrieval of 'enemy' aircraft from Germany and the Occupied Nations. This would be a small 'secret' unit within the Royal Aircraft Establishment – perhaps just half a dozen pilots and a similar number of 'boffins' – and would share captured equipment with the existing Royal Air Force No. 1426 Flight, which had been established at RAF Duxford earlier in the war. The latter had been created to test and evaluate fighters and bombers for air combat, but Eric's job would be to set methodologies for gaining the best information on the technologies which Germany had developed. Eric was particularly keen to see and fly the later model Fw 190 fighters and the new jets, especially the Messerschmitt 262, which was being reported by USAAF bomber crews and whose aura of invincibility had started to be felt by the American crews. If pushed, Eric would always say this small unit was one of his favourites, especially as it gave him access to his second most admired fighter of the war, the Focke-Wulf Fw 190D, the long-nosed 'Dora'.

For a variety of reasons then, Eric was the perfect choice, as he not only spoke German but had also tested many captured German and Italian aircraft during the last few years. His mission of extracting the exotic German designs, about which there was only sketchy intelligence reports, would have the full political backing of Mr Churchill, while he was authorized to strike deals, and achieve the best for the nation, in any way he saw fit. This would include interviewing leading German academics and engineers, taking drawings, notebooks and other literature back to Britain and preventing the UK's allies – American, French and Soviet – from obtaining anything of interest.

As Eric was making preparations, the 21st Army Group, under Field Marshal Montgomery, was moving forward on German soil, with the 2nd (UK) Army given the responsibility of alerting the Farren Mission as to any appropriate technology it found. The first report came as early as 14 April, when British troops overran the airfield at Fassberg, north-east of Hanover, and found two aircraft which 'flew without propellers'. Eric was ecstatic when he read the report. It could only mean jet fighters. The report said that the aircraft were airworthy and had just been flown in from the east by two German pilots evading the Red Army.

Wasting little time, on the afternoon of 14 April, Eric arrived at Fassberg to find two Messerschmitt Me 262 jet fighters, and, as previously reported, both were in flying condition. Eric wanted them to stay that way, so he sought out the local British commander, Brigadier Llewellyn Glyn-Hughes of the Royal Army Medical Corps, to gain permission to select some Luftwaffe prisoners to stand guard. Hughes was impressed with Eric's command of the German language and said it was better than that of his own Intelligence Corps translator. It was then, in granting the permission for Luftwaffe guards, that Hughes asked Eric to spend a day at a special camp, which had just been uncovered by his troops, and where Eric's language skills would be needed to interrogate the guards. Eric was not to know that this was to be one of the most telling parts of his life.

The Bergen-Belsen camp was the first experience of the Nazi death machine for the British Army, and the following Reuters report of the time sums it up:

> One of the British senior medical officers, Brigadier Llewellyn Glyn-Hughes, told the Reuters news agency he saw evidence of cannibalism in the camp.

There were bodies with no flesh on them and the liver, kidneys and heart removed. He said their first priority was to remove the dead bodies from the camp. He was told some 30,000 people had died in the past few months. He said typhus had caused far fewer deaths than starvation. Men and women had tried to keep themselves clean with dregs from coffee cups. Medical supplies were severely limited – there were no vaccines, or drugs and no treatments for lice.

When Eric arrived, he found the camp ringed with troops. The inmates and guards, mainly German and Hungarian, had been left in place due to at least 20,000 cases of typhus fever. Eric was horrified by what he found:

> We thus arrived to find the German and Hungarian troops ready to make a coordinated handover. While this was under way I had a look around, clad in a surgical mask, but quite unprepared for the horrors I was about to witness. There were mounds of dead bodies, most female, all bulldozed grotesquely into pits. Other inmates dressed in prison garb shuffled around like zombies and barely seemed aware of the new activity around them. There were some long huts which I later learnt had been built to house 60 inmates on three-tiered, open slatted bunks, with just

> one toilet for all. They now held 250 dying women in indescribable filth. The stench of these huts has never left my nostrils.

For someone who had, until then, respected the Germans as enemies, and as human beings, Eric was shocked to his core by the conditions at the Bergen-Belsen concentration camp. It was not the death of so many people that affected him so profoundly so much as the suffering, and the inhumane conditions in which the women prisoners, in particular, were kept. It was estimated that from 1943 to 1945 over 120,000 men, women and children were transferred to Belsen, with 50,000 dying of malnutrition, disease and mistreatment by the guards. Hundreds of men were also murdered after being injected with phenol. Among the many victims were Anne Frank and her sister Margot, who both died of typhus a month before the camp's liberation.

In talks in later life, Eric would speak in joyous terms about his pre-war experiences in Germany, and his test flying, but then his face would visibly cloud over when he spoke of the concentration camp and its two commandants, Josef Kramer and Irma Grese, whom he helped to interview thanks to his command of German. Eric recalled that Kramer 'knew the game was up and was ready for death' but the 'loathsome' Grese, known as the 'Beautiful Beast', was a resolute Nazi until she met with the firing squad. 'I can honestly say she was the worst human being I have ever met,' Eric recalled.

It was enough to scar Eric for life. Until his dying day, he would often wake at three in the morning, drenched in sweat, after suffering vivid nightmares. Soon after, Eric would cross paths with one of the men responsible for these horrific crimes.

On 23 May, he was at the British headquarters at Celle, near Hanover, when his friends at Field Intelligence asked for his personal intervention. Earlier in the day, a Sergeant Heinrich Hitzinger of the Special Security Police had been arrested by British troops and brought to the 31st Civilian Interrogation Centre at Kolkshagen camp, Barnstedt. Captain Tom Silvester of the Intelligence Corps felt sure he looked familiar. He did to Eric too. The moment he saw the prisoner walk, he recognized the stooping shuffle of the man he had seen at Berlin and Königswinter before the war. It was Heinrich Himmler, head of the SS, and one of the most wanted men in Germany. Eric said later, 'I think he had a club foot which accounted for his peculiar gait.' Later, as Himmler was being inspected by a doctor, he bit into a hidden potassium cyanide pill and collapsed on to the floor. He was dead within fifteen minutes.

While the experience stayed with him, Eric's time at Belsen had been brief. Within twenty-four hours he returned to the recovery of German aircraft and technology, a job that now began in earnest. Some airfields had already been captured, while others were still in the hands of the Germans – now, since Hitler's suicide, under the command of Admiral Karl Dönitz.

When intelligence alerted Eric to a jet unit at Stavanger-Sola in Norway, this turned into a real find, as Eric recorded finding an array of prized aircraft, such as six Arado Ar 234B jets, at least the same number of Heinkel He 219 night-fighters fitted with the dreaded Schrägemusik upward-firing 30 mm cannon, as well as a host of various early marks of Focke-Wulf Fw 190A single-engine fighters. Most were in good condition, and many were flyable. Eric marvelled that the Germans destroyed or vandalized their older propeller-driven aircraft but left the jets intact, as if 'they wanted to show us what they had achieved despite the bombing and the scarcity of raw materials'. He particularly prized another design from the same stable, the solitary late-model Fw 190D9, nicknamed 'Dora' by German pilots. With its characteristic long nose and 400 mph performance, Eric loved it for its speed and sleek design.

However, while Eric revelled in testing these revolutionary German aircraft, it also caused him some problems. When flying the Dora, Eric was aware that, according to 2nd Army Field Intelligence, all the area to the Baltic coast had been occupied and 'pacified' by the British and Canadian troops of 21st Army Group.

However, the physical proximity of the rump Reich administration in northern Germany meant that some diehard Wehrmacht and Waffen-SS were still fighting.

As Eric approached the area in his Fw 190, still in its Luftwaffe markings, Canadian troops, who were dug in on the west side of the airfield, opened fire, thinking Eric was the enemy. Acting instinctively, he pushed open the throttle and climbed away at over 3,000 feet a minute and hoped that nothing vital had been hit by the infantrymen and their Bren guns. Thankfully, the Dora was quick to respond and evaded the gunfire. This turn of speed convinced Eric that the Dora was just a gnat's whisker behind the Spitfire Mk XIV as the best wartime piston-engined fighter, and more responsive in flight than the American P-51D Mustang.

The pace during these initial weeks was unrelenting. That very afternoon Eric's logbook reveals that he finally flew a Messerschmitt Me 262B, a twin-jet night-fighter: the one he called the 'least weary-looking' on the flight line. Eric's flight was successful, but he knew that if he was to get the best out of the Me 262, he needed expert help.

With this in mind, he spent a morning going through Field Intelligence lists to check the names of German technicians and pilots who had already been detained. He soon came across the familiar name of test pilot Gerd Lindner, No. 8 on the Farren Mission Most Wanted list. Lindner had spent much of the previous two years testing the Me 262, so would be an ideal person to speak to. After a few phone calls, Lindner was made available to Eric to interrogate. He proved both technically competent – and was impressive as a person – happy to talk all things aviation. Eric liked him immediately, and later that month his knowledge of the Me 262 helped Eric fly the single-seat version with more success. Eric could not help but compare it to one of its British rivals at the time, the de Havilland Vampire. While he felt the Vampire 'was a nippy little fighter, with a good turning circle and a useful turn of speed', he felt the British fighter 'would have met its match in the Me 262'.

Out of all the Germans Eric might have hoped to speak to, who could help in mastering an array of machines, Professor Dr Dipl.-Ing. Kurt Waldemar Tank was near the top of the list. Tank had led the design department at Focke-Wulf in Bremen for sixteen years, but being both an engineer and a test pilot made him a rare commodity. Eric had faced Tank's brainchild, the Focke-Wulf Fw 190A design, during the war. He was impressed, and just a little scared of the German fighter. In time he would also fly Tank's strangely shaped yet rather appealing Focke-Wulf Fw 189 Uhu (Eagle Owl), which Eric described as an aeroplane 'offering little in terms of conventional beauty, exoticism or scintillating performance' but which had 'an aura' which defied analysis. Such was Tank's prominence in the industry, the German Air Ministry directed that the chief designer's initials should be used for new aircraft designs, hence the follow-on to the Fw 190D series was the Ta 152.

To Eric's delight, he met with Tank several times. The conversations would have undoubtedly lingered on Tank's Fw 200 Condor airliner design, which as a long-range maritime patrol aircraft had nearly put paid to Eric's flying career, and life, over the Bay of Biscay in October 1941. Tank was clearly a high-value detainee but austerity Britain's newly elected Attlee government could not afford his expertise, nor could it find a place in a shrinking postwar aviation industry for his radical designs. Although Tank was later cleared of any Nazi Party wrongdoing, there was also no room for him in the new West Germany. He did not stay with the reformed Focke-Wulf after the war, deciding instead to go to Argentina, where he worked for Fábrica Militar de Aviones, under the patronage of the dictator and president, Juan Perón. After the latter's fall from grace, Tank went to India to help Hindustan Aeronautics develop its own designs, eventually returning to live in West Berlin in the 1970s. He died in 1983, almost unmarked in the media, and yet Eric rated him supreme among German aircraft designers.

With so much going on all over the country, Eric soon realized that the mission would require a central collection airfield. On behalf of the newly created British Air Force of Occupation, Eric subsequently took personal control of Schleswig airfield, which boasted the Schleswig-See flying boat base, while also being adjacent to an inlet on the Baltic. There was also an added bonus in Leck airfield being nearby, where another prize awaited Eric and the Farren Mission. There Eric's team found a pristine squadron of Heinkel He 162 Volksjäger jet fighters, wrapped in waterproof sheeting to protect their engines. The aeroplane was a simple but distinctive design, with a single turbojet engine carried above the fuselage.

Even at this stage in the war, literally the last few days, the Luftwaffe was still training young fighter pilots to fly and fight in the He 162, which they called the 'Salamander', or sometimes the *Spatz* (sparrow). It never properly entered service, and while it had obvious potential, it was another example of too little, too late. However, it had an innovation which really appealed to Eric: an ejection seat for the pilot.

Although some other German aircraft had been fitted with these life-saving devices, this example particularly impressed Eric. He remembered tales from his youth of the First World War, where Allied pilots were not allowed to carry life-saving parachutes by ground-based generals, whilst their Luftstreitkräfte opponents had them available. Unsurprisingly, many a German life was spared through their use. Eric was therefore particularly happy when Martin-Baker, one of the British aviation companies he most admired, was given access to this German technology. The result was seventy years of the world's best ejection seats, which by the end of 2022 had already saved 7,674 lives in more than 100 countries. This is an example of where all the effort to capture German technology really paid off.

Finding ejection seats and jets was one thing, but the Farren Mission soon uncovered a real treasure trove of German technology, which would change the course of aviation history, as well as Eric's career.

Some flying machines exactly create their purpose in their looks. Those same machines are often forerunners of a new era in aviation. Thus the Messerschmitt Me 163 Komet rocketplane stepped into history when test pilot Heini Dittmar became the fastest man on earth in 1941. The question which has interested aviation historians ever since is did Eric become the only non-German to fly the rocketplane under power?

18

The Komet

On 2 October 1941 during a three-minute flight over the Baltic Sea, Heini Dittmar became the fastest man on earth, travelling faster than 1,000 kilometres per hour. He did so thanks to the revolutionary German Messerschmitt Me 163 rocketplane (aka 'Komet'), which at full power could climb to 30,000 feet in two and a half minutes. On smashing these records, Dittmar gleefully boasted over the airwaves of this new aeroplane being a huge technological advance, and that it might even break the sound barrier. He was not to know that London was listening.

While records no longer exist of the exact content of the radio transmissions, it is known that the report was rushed to London by despatch rider from the listening station at Wrotham in Kent. It said that the Germans had an aeroplane which could fly at *1,000 miles per hour*. This was twice the speed at which Spitfires had been recorded in flight tests.

The transcripts created a panic in the Air Ministry and the technical community. However, the piece of information that created the most fuss turned out to be false. The operators listening on at Hollywood Manor were usually Jewish émigrés who had escaped Hitler's Germany or Austria and could speak fluent German. While they had been trained to listen to German military broadcasts, they weren't necessarily technically trained. Therefore, while the operator listening to Dittmar had heard him use the word 'thousand' when referring to speed, he didn't specify the units. The operator assumed he meant miles per hour, when in fact he was referring to kilometres per hour. In actuality, Dittmar had flown 1,003 km/h (623 mph), rather than 1,000 mph, which would take another ten years to achieve. This was still a quantum leap in performance and far in excess of anything Britain then had in the pipeline.

Thankfully, the small, tailless Komet was only declared operational in May 1944, towards the end of the war. Nevertheless, it was still fearsome opposition. Upon any enemy bombers being sighted, a Komet would be towed to 20,000 feet by two Me110Gs and then released, starting its rocket engines and diving down at the enemy at tremendous speed. The Komet only had enough fuel to do this a handful of times, but such was its pace and performance that it quickly became the stuff of legend.

Indeed, had the Komet been ready a year earlier, it might well have changed the course of the conflict. During the US 8th Army Air Force daylight raids in 1943 it would have been a formidable foe, and the politics of Americans dying in significant numbers over Germany would have likely led to a switch to night bombing, allowing German industry a respite during daylight, instead of being harassed night and day.

When Eric first saw several examples of the Komet, at Husum, by the North Sea coast, he thought it was 'sensational'. He immediately recognized it was the stepping stone to flying faster than the speed of sound and was eager to fly it himself. However, Eric was apprehensive. He knew the fuel that was used was highly volatile and was the cause of a high fatality rate. Over 80 per cent of fatalities in the Komet had been directly attributable to the highly combustible mixture. If Eric was going to get to grips with this fantastic, but dangerous, machine, he realized he needed some help.

Conveniently, several Luftwaffe technical officers were still at Husum and were prepared to help Eric understand the Komet; in fact, they were proud to boast about the machine. Rudolf Opitz, a test pilot, who had been in command, was especially helpful. To avoid fuel issues upon landing, Opitz advised dumping all the fuel beforehand. However, he said that if things should go wrong then bailing out when at 250 mph was impossible, as the hood could not be safely removed at that speed. Worse still, Opitz said that there was also a very high rate of broken backs among those who had flown the Komet, due to heavy stall-in landings. He also spoke at length about the problems of compressibility at high speed, which he had found during two bomber interceptions in 1945, and which Eric, and Farnborough, were still struggling to understand. Indeed, it was said that Eric's hero, Udet, had even died due to this aerodynamic phenomenon. The truth was, however, far less heroic. In 1941, Udet had in fact shot himself, after increasingly poor relations with the Nazi Party. Eric later found out that Werner Mölders, who had helped persuade the budding young Scotsman to fight in the Spanish Civil War, had also died, in a transport aircraft crash on his way to Udet's funeral.

If Eric was going to get to grips with the Komet he was told to first fly the Me 163A training glider, to understand the handling characteristics of the tailless configuration, before a short 'flip' in the Me 163B proper. However, it was quickly apparent that Husum airfield could not be used for such tests due to bomb craters. Therefore, the glider version, the Me163A, was air towed behind a Messerschmitt Bf 110D twin-engined fighter, to the Farren Mission base at Fassberg, with Eric flying a second Bf 110 to watch the Me 163's flight characteristics. This merely whetted his appetite. He couldn't wait to fly it for real.

For now, he would have to make do with flying the Me 163A glider, not under power. This was far from the real thing, but it would at least allow Eric to somewhat get to grips with this unique aircraft. On 26 May 1945, he subsequently logged three flights in the Me 163A glider, when he was towed to 20,000 feet by a Messerschmitt Bf 110. Two weeks later, Eric was towed in an unpowered Me 163B to 32,000 feet, gliding back to Farnborough. Even without the rocket motor Eric found it a difficult machine to handle. It was 'generally unstable', and had to be 'firmly controlled all the time'. Indeed, on 10 October 1946, he almost crashed an Me 163B that had been towed to 16,000 feet by a Spitfire Mk IX. He also found landing the glider a challenge, as the view was obscured. If this was difficult to operate as a glider, then Eric had no idea how he would fare when it was under power. Still, he was desperate to try. But there was a problem.

Powered flight in the Me 163 was banned by the Royal Aircraft Establishment's director, as the use of the German T-Stoff concentrated hydrogen peroxide in the fuel was prohibited in Britain. All tests were therefore to be towed flights only, with any powered flight in the Komet needing to be authorized. For now, there looked to be little opportunity to fly the aeroplane under power. Until then, Eric could only satisfy himself by learning as much about the Komet as possible.

He decided to track down those German inventors, scientists and pilots who had been responsible for the aeroplane's development. If they could be persuaded to cooperate then their information would prove to be a veritable goldmine. However, while finding some of these men and women was easy enough, getting them to cooperate, and keeping them alive, was a whole other matter.

When author Frederick Forsyth wrote the acclaimed thriller *The ODESSA File*, it was a work of fiction but also based on very hard fact. The ODESSA organization in the story was in fact a real Nazi hardcore underground organization, which vowed to take revenge on any Germans, including civilians, who cooperated with the Allies. Such betrayal was seen as high treason. Despite this warning, several brave Luftwaffe volunteers helped the Farren Mission, and without their courage the collection of German aircraft and technology would not have been possible. For these reasons, when Eric soon came across someone of immense value, he had to tread very carefully.

In early May, 30 Assault Unit, a specialist commando unit which was the brainchild of Commander Ian Fleming, of later James Bond fame, in Naval Intelligence captured the dockyard and works in Kiel at the Walterwerke, or, to give it its full name, Hellmuth Walter Kommanditgesellschaft. Here they found rocket motors, the likes of which were used in the Komet, as well as research into air-independent propulsion for submarines; both technologies far in advance of those of the Allies.

Perhaps the most important find of all was the scientists responsible for this technology. Dr Hellmuth Walter, and his technical assistant Professor Dr Kramer, both agreed to talk, and gave up their secrets as to how Walter had designed, engineered and concocted jet engines, rockets, high-speed boats, jet-propelled tanks, long-range projectiles and guided weapons, as well as the rocket motor used in the Komet. To Eric's delight, Walter even put on a demonstration for him.

Stood behind a 9-inch-thick glass panel, Eric watched on as Walter and his team, dressed in rubber aprons, gumboots and hats, switched on the Komet's motor. Eric couldn't believe its power. The resulting roar was so loud that it shook the room. It was a 'frightening' display. Yet Walter wasn't done.

Next, Walter took Eric to a lab, where he picked up two glass rods. On one was a droplet of hydrogen peroxide, and on the other a solution of hydrazine hydrate in methanol. This was the fuel that was used to power the Komet, known as *T-Stoff* and *C-Stoff* respectively, which, when mixed, was explosive enough to propel the 2 ton Me 163 at nearly 600 mph at altitude. Walter dripped one drop on to the floor, then another on top of it. On impact, such was the 'violent explosion' that it blew the rods out of Walter's hands. Eric was stunned. He truly realized now just how dangerous the Komet could be, and yet he became ever more desperate to fly it under power. As he said himself, such was the aeroplane's allure, it was as if 'she was the mystical siren on the Lorelei Rock in the River Rhine that lured mariners to destruction'.

While finding the technology behind the Komet was vital, as well as the men who invented the engine and fuel itself, it was equally crucial to track down the man who had actually designed the tailless aircraft. That man was Alexander Lippisch,[1] who had first worked on tailless designs for the Zeppelin company, as well as gliders, before he created delta wings for rocket-powered aircraft at Messerschmitt in Augsburg in 1939. Thanks to Walter's motors, and Lippisch's design, in 1941, at the military aviation test facilities at Peenemünde-West, Lippisch's team successfully flew the Me 163 for the first time.

However, Lippisch, always the engineer, and Willy Messerschmitt, the Nazi Party member, could not agree about the development of the Me 163, and the Lippisch team was subsequently transferred to the Austrian aeronautical research institute in Vienna. There they were to work on the theoretical results from a wind tunnel test, which indicated that the delta-wing format was the quickest way to achieving supersonic flight. It was

1 Lippisch supervised the PhD thesis of Beverley Shenstone, in which the latter created the elliptical wing which was so important in the development of the Spitfire.

thought that a Lippisch design, the P 13a delta, might be the right vehicle if a ramjet could be used to power it to supersonic speed; it would still need a turbojet for take-off. By 17 May 1943, Lippisch also had a delta-wing, turbojet-powered fighter on the drawing board. It even reached the wooden mock-up stage in 1945, with only the lack of powerful turbojets, and the end of the war, stopping it from becoming a reality.

With Lippisch captured by the Allies, Eric was able to secure an interview with him at Kransberg Castle, high in the Taunus mountains of the German state of Hesse, north of Frankfurt. This was to have been the headquarters for Hitler's invasion of Britain, Operation Sealion, and was subsequently used by the Luftwaffe as a personal retreat for Hermann Göring, and later as a recuperation hospital for some of the wounded from the Russian Front.

Notes of what was discussed at the meeting were sadly not kept, but undoubtedly Eric would have gleaned useful information about the Komet and its flight characteristics.

Lippisch was another German designer whom Eric believed could be of enormous use to Britain. He was, however, left frustrated, as, rather than work for the British, Alexander Lippisch and several of his family members instead accepted offers to work for the US government. At the first opportunity, he was therefore shipped off to the White Sands Missile Range in New Mexico. Still, Lippisch and Eric continued to develop a relationship, and the Lippisch family even maintained contact over the succeeding decades.

Although men such as Walter and Lippisch were people of the utmost importance to be found and interviewed, others came as more of a surprise. At No. 5 on the Farren Most Wanted list was Hanna Reitsch. She was arguably Germany's most famous wartime female pilot, and Eric, of course, had a special connection to her, having spent time with her in Germany in those halcyon days before the war.

Since they had last met, Reitsch had been working at the edges of technology, firstly with the Messerschmitt bureau, on the giant Me 321 transport glider, which had been designed for the aborted invasion of Britain. Yet of most interest to Eric was Reitsch's work with the Komet, particularly her time on the test team at Rechlin, near Berlin, involved in the development of the 'flying bomb' terror weapon, including flying a crewed version and the Me 163 rocketplane.

Reitsch showed incredible bravery testing such a volatile and dangerous machine to the limits. In fact, it almost killed her. On 30 October 1942 she crashed, breaking numerous bones and suffering multiple fractures to the skull. Nevertheless, before she passed out, she still managed to write a quick report about the undercarriage release problem that had caused severe handling issues. It was a great act of courage, which, in November 1942, resulted in the award of the Iron Cross First Class; Reitsch was the only woman to be so honoured. The pictures of her with Adolf Hitler tell their own story of her pride and joy at being recognized in this way by her Führer. Her time in hospital included reconstructive surgery on her nose, but after some flying practice to regain her confidence in the air, she was back in the Komet and took part in more trials. She was still hoping to fly the rocketplane under power but to her disappointment was now considered too valuable as a publicity asset to be risked in another crash.

There was no doubting Reitsch's courage, but sadly nor was there any doubt about her allegiance to the Third Reich. Towards the end, she even toyed with the idea of defending the Reich against American bombers by creating a squadron of suicide fighters, suggesting that she should lead the first attack mission. Had it not been Flugkapitän Hanna Reitsch who had mooted the idea, such actions would have been seen as defeatist and subject to some official sanction. Somewhat ironically, considering his eventual demise, Hitler told her personally that it was 'un-Germanic'.

In April 1945, Reitsch's dogged loyalty to the Third Reich, the dream of a reforged Germany and her talent for flying all led her towards one of the most infamous places in history: Hitler's bunker.

While Reitsch was visiting her family in Salzburg, she received an urgent call from Generaloberst Robert Ritter von Greim, believed to be her lover, who needed her to urgently fly him to Rechlin. The reasons for doing so were unclear. Nevertheless, as ordered, on 25 April she set out with von Greim in a Junkers Ju 188 twin-engined bomber, skirting known Red Army positions, using landmarks which she had already committed to memory from previous flights at low level.

At Rechlin, von Greim was finally briefed on his operation. He was to fly to Berlin to meet with the Führer at the *Führerbunker*. Reitsch was determined to go as well. However, it seemed Gatow airfield was the only viable landing strip in Berlin, but even that was difficult to reach due to the advancing Red Army. While they had initially intended to fly via helicopter, this had since been destroyed in an air attack. The only aircraft available was a Focke-Wulf Fw 190 two-seat trainer. A young *Feldwebel* (sergeant), who had experience of flying the aircraft, was subsequently ordered to fly von Griem to Gatow. Although there were only two seats available, Reitsch wasn't going to miss out and somehow squashed herself into the rear fuselage's radio compartment.

In the early afternoon of 26 April, following an escort of at least a dozen Fw 190A-5 fighters, the Focke-Wulf landed safely at Gatow. From there, von Greim and Reitsch transferred to a Fieseler Storch, the same one which is believed to have taken Albert Speer, the armaments minister, in and out of Berlin a few days before. As dusk fell, the Storch set out for a makeshift runway, which had been constructed by sappers in the Tiergarten and was located near the Brandenburg Gate, the closest landing zone to the Reich Chancellery. On the way, the aeroplane was shot at over the Grunewald, and von Greim was hit in the foot. He passed out through loss of blood, forcing Reitsch to lean over him in order to land. This was a feat at the best of times, yet in failing light, and under fire, it was nothing short of superb airmanship. The Storch did not fare so well and was destroyed later in the day as the Red Army closed in.

In the *Führerbunker*, while von Greim was treated for his wounds, he was promoted to *Generalfeldmarschall* and given the title *Oberbefehlshaber der Luftwaffe*. He was also instructed by Hitler to fly out of Berlin, find Heinrich Himmler and put him under arrest for treason. Hitler's doctor, Theodor Morell, also gave von Greim and Reitsch cyanide pills, to take in case they were captured. Reitsch wanted to stay in the bunker with Hitler, but the Führer refused. In a postwar interview, Reitsch said, 'It was the blackest day when we [she and von Greim] could not die at our Führer's side. We should all kneel down in reverence and prayer before the altar of the Fatherland.'

Taking the cyanide pills with them, von Greim and Reitsch, in an Arado Ar 96 trainer, flew back to Rechlin, having to avoid the anti-aircraft fire from the Soviet 3rd Shock Army, who put up a sustained barrage because they thought Hitler was on board and trying to escape. The Ar 96 arrived in Rechlin unscathed, and, after refuelling, von Greim and Reitsch flew on to Plön, where Admiral Karl Dönitz had set up his headquarters after Hitler's suicide, briefly taking over as German head of state. Yet events were now impossible to escape. As Germany fell, they separated; von Greim also took his own life, and Reitsch fled. She was nowhere to be found.

For all these reasons, Eric was keen to find his old friend. She had a very rich seam of high-grade intelligence that could help Britain in the immediate postwar aeronautics race for civilian, as well as military, flight. This was what the Farren Mission was all about. Moreover, she also might have vital information he could use to fly the Komet.

The day after surrender, Eric received information that Reitsch might be in a hospital, near Kitzbühel, supposedly after suffering a heart attack. On 14 July, Eric arrived at Leopoldskron Castle near Salzburg, where Reitsch was in captivity and recuperating. As Eric walked in, he immediately recognized her as the aviation idol he had met before the war. However, she was in a 'very emotional state'. Not only had Germany lost the war, but her father had committed suicide, not before shooting her mother, her sister, the grandchildren and even the maidservant. Everything she had held dear in the world had turned to dust.

In a three-hour meeting – it could not be called an interrogation – Eric found that Reitsch spoke freely and recounted her test flying. It may be that she felt able to share this information because she recognized Eric from Berlin in 1938 and their time together with Ernst Udet. She was, however, in such a highly charged state that at times Eric had to slow her down. While they mostly talked aviation and patriotism, when it came to Eric explaining his recent experiences at the liberation of the Bergen-Belsen death camp, she said that she did not believe him. 'Her devotion to Hitler and a form of German nationalism made my blood run cold,' Eric admitted.

Yet what Eric really wanted to know about was Reitsch's experience with the Komet. While Reitsch relayed her experience of testing the Komet while gliding, she confided that she had never flown it under power. Eric found this extremely disappointing. Reitsch was one of the pilots he respected above all others and he was relying on her knowledge to fly the Komet himself. Although Eric left somewhat empty-handed, his relationship with Reitsch didn't end here.

After the war, Reitsch moved to Ghana to set up

a gliding school and occasionally kept in contact with Eric. In September 1971, they met once more at the first International Helicopter Championships in Bückeburg. They exchanged letters for a while afterwards, including when she became the first woman admitted to the Society of Experimental Test Pilots, before she passed away on 24 August 1979. Eric recalled their last telephone conversation on 23 August, the day before her death, and a letter she wrote to him at about the same time. In the letter she told him that 'what began in the bunker must now end that way'. Eric took that to mean that she would finally take the cyanide pill given to her by Hitler's doctor, Theodor Morell, in the Berlin Bunker just two days before the Führer's suicide on 30 April 1945. The letter is no longer extant and was sent to Reitsch's brother for the family to read. Unusually, Eric did not keep a copy.

At face value the relationship between Reitsch and Eric still puzzles. How could a British fighter pilot and a fanatical Nazi remain close through everything that had happened during the war? Yet Eric would often speak fondly of Reitsch and of the times their paths crossed. He had found her 'proud and energetic' when they first met but later thought her too committed, with a dogged loyalty to Hitler and Germany.

After he died in 2016, Eric's papers revealed a vast collection of notes, newspaper and magazine cuttings, as well as photographs of Reitsch, particularly pre-war. The inescapable conclusion is that Eric had developed some form of fascination, even an infatuation, for her during the late 1930s. If so, it wasn't reciprocated because she was more interested in flying than in men. In any event, it seems that Eric could forgive her fanatical devotion to the Führer because of her skill in the air.

While Reitsch had little valuable information Eric could use for the Komet, there was another name on his wanted list who was certainly of interest. Hermann Göring was the infamous commander of the Luftwaffe, and Eric almost missed him, as he was nearly executed by his own side.

On 30 April, on the orders of Adolf Hitler, Göring was arrested by the Gestapo and held under house arrest on the charge of treason, before being freed by a passing Luftwaffe contingent on 5 May. Göring decided to head to Austria with them, intending to reach 'the castle of my youth' at Burg Mauterndorf, to which he had fled in 1923 after the failed 'Beer Hall Putsch'. However, with the Soviets advancing through Austria, Göring was forced to change his plans. At Bruck, near Radstadt in the Salzburg region, he came across elements of 36th (US) Infantry Division. In the early hours of 6 May 1945, he had little choice but to surrender to the American soldiers. He was subsequently taken by air to Mondorf-les-Bains in Luxembourg, where the Palace Hotel had become a holding prison for notable Germans, dubbed 'Camp Ashcan' by the US forces.

Eric was desperate to speak to Göring, but he was heavily guarded by the Americans. However, on 16 June 1945, he found an opportunity. Having successfully acquired all of the airworthy Arado Ar 234B jet bombers in Denmark and Germany for the British, Eric discovered that Colonel Watson's team did not have one for the USAAF to study. Knowing they were keen to have an example to take to America, Eric traded an Arado for an hour with Göring, which later, it is said, became two hours for two airworthy bombers.

The US Army lawyer assigned to him gave Eric strict instructions to stay away from any political questions, but Eric simply wanted to talk about the Luftwaffe, aircraft and technology. 'I was very uncertain as to how far I was going to be allowed to go with him,' Eric remembered. Thankfully, while being weaned off drugs and alcohol, Göring was also subjected to psychological 'encouragement' to talk.

When Eric entered the cell, Göring did not look up at first but remained 'almost supine, certainly docile' – so much so that Eric thought he had been sedated. That changed when Eric spoke to him in German and explained that he was a pilot, a test pilot, and former fighter pilot no less. The German beamed and started talking freely about his time as the head of the German

Eric and Lynn were married in Belfast on 17 January 1942. The relationship would last for more than fifty years. Eric was on survivor's leave for the wedding, and is wearing the ribbon of the Distinguished Service Cross won in the Battle of the Atlantic.

During radio trials in 1943, an engine failure forced Eric to land through invasion obstacles – 'a really shaky do'.

When test flying the Seafire in Scotland, Eric looped through the spans of the Forth Bridge. He was spotted, but because the naval Spitfire was still secret, Royal Air Force units in the locality were blamed.

When Eric landed a Mosquito on HMS *Indefatigable* on 25 March 1944, he flew into the record books and secured his reputation as Britain's greatest trials pilot.

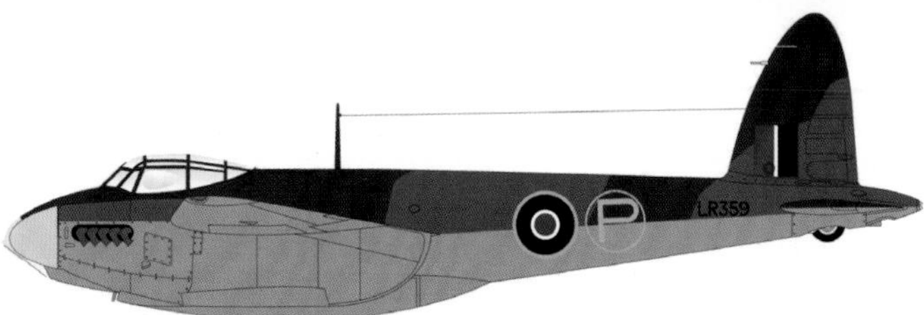

Eric carried out the first deck landing of a twin-engined aeroplane in this Mosquito, which had been modified by Marshall of Cambridge on behalf of de Havilland and the Air Ministry.

Eric bailing out of a Hawker Tempest during trials at Farnborough in 1944 into finding ways of defeating the German V1 flying bomb.

Eric used 'his' Bell Airacobra for deck-landing trials, making him the first pilot to land and launch from an aircraft carrier in a tricycle aeroplane on 22 June 1944.

Rocket launch trials at Farnborough nearly went wrong in 1944 when Eric was testing a Seafire Mk IIc and there was a trolley malfunction caused by human error. Eric, of course, managed the emergency with aplomb.

The Nazi death camp at Bergen-Belsen, near Celle, was just one of dozens of camps liberated by the Allies in 1945. It was appalling and Eric never forgot the stench and misery.

Commandant of the female inmates at Bergen-Belsen, Irma Grese. Memories of meeting her haunted Eric for the rest of his life.

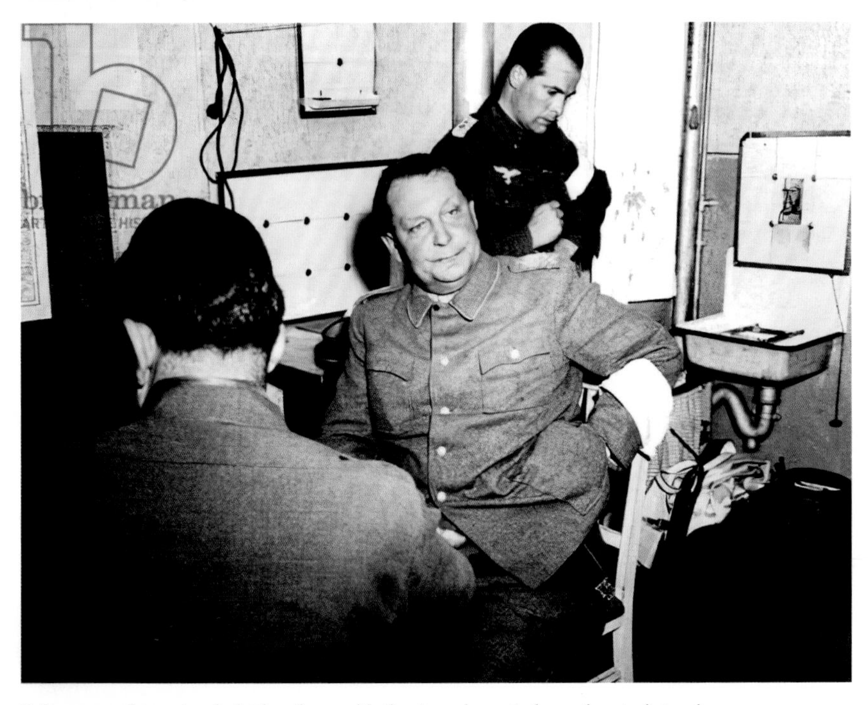

Eric swapped two Arado jet bombers with the Americans to have time to interview the head of the Luftwaffe and former German fighter ace, Hermann Göring.

In September 1944, British troops found an airworthy Heinkel He 177 bomber at Bordeaux. Eric flew the five-hour ferry flight back to Farnborough. His verdict: 'not a good ship'.

William Farren, Director of the Royal Aircraft Establishment since 1941, led the mission to capture German technology.

The unconventional twin-engined Dorner D 335 proved troublesome and crashed in 1946, killing one of Eric's fellow test pilots.

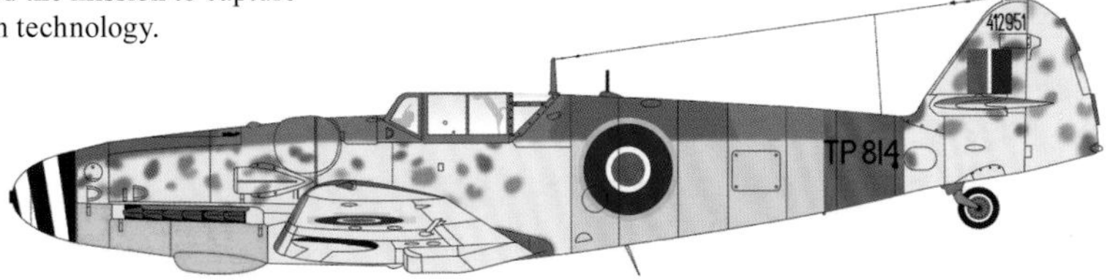

Suitably re-marked in British colours as part of RAE's Enemy Aircraft Flight, this Messerschmitt Bf 109G-6 was flown by Eric in 1944.

An early German jet was the Heinkel He 162, assigned to the RAE Enemy Aircraft Flight for trials, including those of ejection seats. Eric flew this particular one on 30 November 1945.

Pictured at Lubeck in late 1945, the two most successful German fighters of the Second World War – the Me 262 jet, one of Eric's all-time favourite aeroplanes, and the Fw 190 fighter.

This Me 163B was captured and used for trials by Eric as part of the Farren Mission. It was later returned to Germany thanks to Eric's intervention.

Eric received this postcard from Berlin Gatow Museum in 1988 in thanks for his help persuading the UK Ministry of Defence to release the rocketplane for display in Germany.

Eric was very cagey about flying the Me 163 rocketplane under power – as his crash at RAF Wittering on 15 November 1945 proved, it could bite.

Eric flew into the history books on 3 December 1945 by landing a jet aircraft on an aircraft carrier. At 11:28 a.m. he brought Vampire LZ551/G to HMS *Ocean* in the English Channel.

The first jet landing was a significant event in naval aviation and paved the way for more trials in Britain and America. These were the days before 'bone domes', as test pilots dressed little differently from their forebears a generation before.

Eric lands a Sikorsky Gadfly on an aircraft carrier, probably HMS *Ocean*. He was demonstrating the versatility of helicopters for Fleet use.

If Eric had a real nemesis in the test flying world, it was Chuck Yeager. They met at Edwards Air Force Base in 1951, and there was mutual distrust from the off. Yeager was the first to officially break the sound barrier and that really hurt Eric's pride.

Based on German research, the de Haviland DH 108 Swallow was designed to break the sound barrier. Eric reckoned he was lucky to have flown it and lived.

The 'Prone Pilot' Meteor was created with Eric's assistance during a series of trials investigating how to protect pilots breaking the sound barrier. Eric did not fly the machine in 1954, but he always claimed 'a certain ownership'.

On 12 August 1949, Eric flew the ill-fated Saro SR/A1 flying boat jet fighter. It ended in a crash from which Eric was lucky to have survived.

air force. He complained about the lack of resources, the wonderful developments of the German Reich and being betrayed by an international Bolshevik-Jewish conspiracy. Eric ignored the rhetoric and focused on why he was there. 'We talked about the [Me] 262 and [General der Jagdflieger Adolf] Galland's view that Hitler hadn't used it properly,' Eric later remembered.

Göring was completely loyal to Hitler, even at this stage, and tended to blame Galland for the problems of air defence against the Allied bomber offensive in 1944 and 1945. He insisted, 'It was nothing to do with Hitler's direction on the Me 262 being converted as a bomber instead of being exploited as a fighter.' Göring also spoke about the issues with the structural integrity of the Bf 109 fighter, which Galland wanted to be made lighter to give a better rate of climb, and which resulted in 'the wings coming off in combat'. Göring told Eric that Hitler remained loyal to Galland to the end and would not allow the fighter ace to be court-martialled for a list of misdemeanours, which Göring had identified or made up. Ironically, Eric and Galland went on a lecture tour of America together before the latter's death in 1996. 'Galland was a mixed blessing,' Eric admitted, 'I didn't particularly take to him.' The relationship had all the hallmarks of a personality clash.

With his time running out, the topic of the Battle of Britain came up. 'It was a draw,' Göring told him.

'We withdrew to regroup for the Russian campaign. Certainly, our attrition rates were high, but so were yours.' Eric did not argue. Unfortunately for Eric, Göring had little to share when it came to the Komet.

On rising to leave he was confronted with the German doing the same and offering his hand. He realized that the Americans, neither of whom present could speak German, would intervene if he shook hands, causing embarrassment all round. So instead, Eric stood up and smiled, offering instead of a handshake the First World War pilot's salute to a comrade, taught to him in 1936 by Ernst Udet: 'Hals- und Beinbruch'. Eric found out later that the greeting has its origins in Yiddish. It showed, Eric thought, that the Luftwaffe in general, and Udet in particular, were not tied to the Nazi dogma and would not have minded, perhaps even rejoiced in the irony.

Eric left feeling vindicated in his trading of jets for time with Göring and would always say that it was the high point of the interrogations he carried out. He did not like the man, nor was he impressed by his record, but felt that he was a fighter pilot caught up in the politics of greed. He was certainly one of the most fascinating men of the twentieth century. Göring stayed in Luxembourg until September 1945, when he was flown to Nuremberg, where the Allies were setting up the Military Tribunals. Although condemned for guilt 'unique in its enormity', he avoided the death penalty. Despite this, he committed suicide on 15 October 1946.

Time was now running out if Eric was ever to fly the Komet under power. In his words, it had become an 'obsession'. While it was still strictly prohibited for him to do so, he began to devise a clandestine operation. He knew that there were some Komets available at Husum airfield, but these would soon be transferred back to Farnborough, where the special rocket fuel would be disposed of. In the circumstances, it was therefore now or never.

At Husum, Eric informed the ground crew that he was going to fly the Komet under power. Everyone was apprehensive about this, not only because it was illegal, but also because it was still considered highly dangerous. But Eric was undeterred. He had flown the Komet as a glider and had spoken to a wide variety of people who had been involved in its creation and testing. While the crew were apprehensive about what would happen if something went wrong, they decided to put their trust in Eric and keep his proposed flight secret.

At 6 a.m. the following morning, Eric climbed into the cockpit, fired up the engine, signalled to the ground crew he was ready and then set off down the runway with an explosive burst, which pinned him back to his seat. At 725 kph Eric reached 32,000 feet in just under three minutes, leaving a plume of smoke behind him. In this 'runaway train' Eric proceeded to practise a number of dives, staggered at the Komet's ferocity, yet unsure if it was stable enough yet to enter into combat with a fighter. With the rocket fuel soon running out, Eric glided the Komet gently back to Husum. It had been the flight of a lifetime, one Eric would always remember fondly.

But was any of this true? It was only in later life, and in later editions of his autobiography, that Eric revealed he had flown the Komet under power. Some therefore felt that it was another example of Eric's tendency to burnish his records with stories that didn't quite stand up to scrutiny: the orphan, still conscious about his background, trying to stay one step ahead.

Five decades later, at a Farnborough Air Sciences Trust fundraising dinner, a guest asked Eric if he had really flown the Komet under power, as he had long contended. The questioner was Richard Brown, whose late father, George Brown, had been part of Eric's ground crew at the Royal Aircraft Establishment. Brown was at Farnborough when the Me 163B was towed behind a Whitley bomber for gliding trials in June 1945. Crucially, he also worked as Eric's ground crew when they decided, among themselves, to try for a powered flight at Husum. Richard Brown wanted confirmation of what his, by then late, father had told him about what had occurred that day. Eric confirmed that it was all true and when asked why he had kept the flight a secret, he was direct in his response: 'Court martial – if I had recorded it against a direct order – so I left it vague and didn't talk about it until 2006.' This seems a satisfactory excuse, and the fact that other witnesses have verified Eric's version of events leads me to believe that he did indeed fly the Komet under power. Yet while he suffered no issues flying it under power, he was soon to have a very different experience flying it while gliding.

On 15 November, while Eric was at Wittering for skid tests on the Komet, he nearly flew his last test – ever. A little fast on touch-down, the skid collapsed, jamming him in the cockpit and causing the aeroplane to veer across the airfield for more than 600 yards until it almost hit the boundary fence. 'I had a fair number of cuts and abrasions,' he reported later, 'and it took a while for the emergency team to cut me from the wreckage.' It appeared his most serious injury was a heavily bruised spine, but later it was found he had a hairline fracture of the neck.

While Eric was fascinated by the machine, he admitted that British and American engineers/technologists needed to understand its many shortcomings if it was to pave the way for a supersonic fleet. However, although the Americans had acquired some Me 163 airframes and were keen to use them to gain information on tailless flight for their own bid to break the sound barrier, they were not successful in flying one, even as a glider, as Eric recalled:

> My opposite number was collecting [one of] these aircraft – I had a very good relationship with him, and we operated closely together. He took quite a number of 163s back to the US. They attempted to fly them off Edwards Air Force Base, or some such. They did three attempts, and the tow rope broke every time. I got a message from them saying could I send the specification of our tow rope. We were towing [the Me 163] with a Spitfire Mk IX when we did the tow trials. We sent them [the specification], but whether they used it, I don't know. I doubt it, because it wouldn't have broken, but they admitted they'd never flown the 163.

By September 1945, the Farren Mission was being wound down and was regarded as a success. Seven supersonic wind tunnels had been found, broken down and collected, while twenty-six top scientists and engineers were invited to move, with their families, to Britain. One of those German scientists was Dietrich Küchemann, a leading exponent of supersonic flight and swept-wing theory. His work on tailless aeroplanes remains relevant today, and Eric would say he was ahead of his time. 'He once told me in 1946 that he was thinking about a tailless supersonic airliner – I thought that was unbelievable, but the result was Concorde just twenty years later.'

As if high-speed delta-wing performance was not enough, by 1960, Küchemann had given Britain the lead in fly-by-wire technology, which did away with cables and pulleys for controlling the moving surfaces such as the elevators and ailerons of an aeroplane. This use of electronic pulses revolutionized military and civilian jet aircraft handling, making it safer and more responsive. Küchemann lived in the Farnborough area for the rest of his life and was awarded the CBE and gained membership of both the Royal Society and the Royal Aeronautical Society as a fellow.

The results of America's Operation Paperclip were to be rather more spectacular. The arrogant Nazi Dr Ing Werner Magnus Maximilian Freiherr von Braun, the designer of the V2 ballistic missile, found his past crimes overlooked and he was relocated to Texas. There, he developed a series of space rockets, including the mighty Saturn V booster, which would eventually take NASA astronauts to the moon in 1969.

In the years following the war, sightings of UFOs also exploded across America, particularly in Arizona. Some might have felt that little green men had arrived on earth, but those involved with the Farren Mission, like Eric, suspected something rather different. In Germany, the famed Horten brothers had been developing the Parabel (Parabola) project, which had the possible designation of Horten Ho IX. This concerned aircraft that resembled so-called 'flying saucers' and was unlike anything seen in the skies before. Following the war, the Horten brothers had moved to Arizona, where they helped the Americans research and develop these aircraft. It has therefore been speculated that these so-called UFOs were actually the work of the Hortens.

With the Farren Mission complete, Eric's findings in Germany, not least his experience as the only non-German to fly the Komet rocketplane, would lead him to his next great aviation adventure, the holy grail of test flying in Britain and America: breaking the sound barrier. But the question was: who would break it first? The race was on.

Eric cherished the dream of becoming the first man through the illusive sound barrier – flying faster than the speed of sound. He worked day and night with the boffins at Farnborough and at Miles Aircraft to develop the M52, including having the cockpit designed around him. This extraordinary picture shows the cockpit test vehicle at Farnborough in about 1946.

19

The Sound Barrier

There were many in Britain who had thought that supersonic flight was impossible. Sceptics claimed that no aeroplane could be built with sufficient structured strength to stand up to the aerodynamic forces involved, nor was there a powerful enough engine to propel it to such speeds. Yet by 1943 intelligence indicated the Germans had already flown a ram-jet and a jet-powered fighter, while the Komet would also soon cause a stir. With this in mind, breaking the sound barrier no longer seemed such a far-fetched idea.

On 4 May 1943, the Government Chief Scientific Officer, Lord Cherwell, created a Top Secret Supersonic Committee under Dr Benjamin Lockspeiser, the deputy director of the Royal Aircraft Establishment Farnborough. After looking at the data, Lockspeiser persuaded the Air Ministry to raise secret specification E24/43 for an airframe and engine which could climb to 36,000 feet in 1.5 minutes and reach speeds of 1,000 mph. The contract was given to Miles Aircraft, where Lockspeiser's son David, later a talented aircraft designer, was an apprentice. The design, given the working name of M52, would feature a Power Jets turbojet engine from the design board of Group Captain Frank Whittle, while it also had two revolutionary features: an all-moving tail-plane and extra-thin wings. These wings were so sharp that they became known as Gillette wings, with the test aeroplane, modified by Miles Aircraft, known as the 'Gillette Falcon'.

A little later, in 1944, the Committee was told that if the calculations were correct, the E24/43 should reach 645 mph at sea level. This looked to be a staggering breakthrough. However, the E24/43 transonic research programme was facing problems. For example, it was found necessary to reduce the operating height by 10,000 feet to 50,000 feet for aerodynamic reasons. In the meantime, the weight had increased, making the possibility of achieving supersonic flight, even a dive, a real issue. Uprating the propulsion system of the E24/43 with turbojet and rocket motors was even discussed in June 1944 but abandoned as too complex and, presumably, costly. Still, the research continued.

At this time, Eric and Squadron Leader Tony Martindale were involved in high-speed trials of their own at Farnborough. Throughout 1944 and 1945 the Farnborough team acquired several Mk XI Spitfires, the fastest production fighter then available, to research the potential of supersonic flight. The beauty of the Mk XI was in its aerodynamic profile, which, without gun ports or ammunition blisters, presented a clean face to the airflow. When the Mk XI arrived at Farnborough in 1944 it was also modified to carry an 'automatic observer' recording system, which recorded everything that the pilot did in terms of control inputs.

The aim of the trials was simple. Martindale and Eric were to put the Spitfire into a dive and get as close to Mach 1 as was possible, all while recording the effects. In doing so, Eric took the Spitfire to Mach 0.88, but in recovering from a transonic dive at this speed he reached his own physical limit:

> In transonic testing, when the compressibility effect gets into the game, you find that you usually get a violent nose-down change of trim and you have to counteract it unless you want to lose control, you have to hold the Mach number constant once you've got to lock off the throttle and use two hands on the stick. It's as heavy as that. My limit was about 100 lb pulling.

As you might imagine, this was all tremendously dangerous. On 27 April, Eric witnessed, and monitored by radio, Martindale taking the Spitfire to high altitude, about 41,000 feet, and calling, 'I'm diving now.' Following this, he said nothing more. This wasn't unusual, as he would have then been concentrating on the manoeuvre. But in this case, something had gone badly wrong.

The first Eric knew that something out of the ordinary had occurred was Martindale's shaky voice on the radio saying, 'I've just had something happen . . . but I'm back here at 42,000 feet. I can't see out of the aeroplane because of oil on the windscreen, but it feels all right. I haven't got any power, but it feels like I can manoeuvre it.' There was no power because there was no propeller! In the dive, the airscrew (spinner, propeller and gearing) had come off, with the engine gears also departing from the airframe, and the Spitfire becoming tail heavy as a result. The theoretic g-limit of the Spitfire was 10 g, but in this instance it had been 11 g. Eric recorded that Martindale had needed 'all his strength', measured at over 100 lbs, to pull the Spitfire out of the speed dive before blacking out.

As Eric waited for Martindale to issue a further report, still not aware there was a major problem, the Squadron Leader came back on the radio. He said that he could see Farnborough by looking out of the side of the cockpit and could crab the Spitfire back there by lowering the undercarriage on 'emergency' and using the compressed air system designed for the eventuality of losing all hydraulic pressure.

Approaching 3,000 feet, Martindale pulled the handle for the undercarriage. 'It's worked,' he told Eric. Soon after he was able to land safely back at Farnborough, with the fighter, and himself, intact. It was a remarkable feat of flying. As Eric said later, 'It should have broken up by that time, and when [he] landed I went out to meet him. Actually at the leading edge, where the wing meets the fuselage, there was a gap of about 3 inches on each side.' Such was the pressure that the wings had become very slightly swept back.

Eric always thought that more attention should have been paid to the swept wings. Indeed, the world had already failed to take notice of the work by Adolf Busemann from the University of Dresden, who had delivered a paper on swept-wing flight as early as 1935 at the Volta Congress, organized by Italian dictator Mussolini. He postulated that a swept wing would allow a smooth speed transition from subsonic to supersonic, as if there were no sound barrier. In time, he was to be proved right.

Two months later, Martindale was up again in a Spitfire Mk XI. This time the airscrew did not leave the airframe, but instead went into overspeed. The engine subsequently failed, and he was forced to glide down into a Surrey field and dive under some power lines. He was badly hurt, and broke his back, but retrieved the auto-recorder before he blacked out from the pain. Eric was again a witness to the flight and remembered, 'He hadn't gone quite as fast as on the previous occasion, but he'd gone to 8.95 g or something like it. He was a very good pilot.' High praise indeed.

While Eric and Martindale were risking their lives

trying to reach Mach 1 in a Mk XI Spitfire, the design for the M52 was almost complete. The project now required a test pilot, who might have the honour of being the first man to break the sound barrier. The natural choice was Martindale, but his 15 stone, 6 feet-tall physique ruled him out as being too large for the proposed cockpit. All eyes subsequently turned to Eric. He not only had all of the required experience, but he was also just 10 stone in weight and 5 feet 6 inches in height. Martindale's loss was very much Eric's gain. On 4 August 1944, Eric received a short, simple and classified note from the RAE director, William Farren, suggesting that he take a 'close interest in the supersonic research programme'. He was delighted. His previous accomplishments had made his name in the aviation community, but this promised to make him known far and wide. It was truly the chance of a lifetime.

Soon after, the RAE designed a capsule with which Eric was intimately involved: 'I sat in the seat dozens of times while they moulded things around me . . . it was a reclining seat, and they wanted to make sure I could reach the rudder pedals and the various instruments in a climbing position.' This capsule was very important. Such were the clear dangers involved in testing for supersonic flight that the means of pilot escape, in the event of a structural or other failure, was a prominent issue.

In the meantime, Eric also started to work with Dr Ben Lockspeiser, the chairman of the Supersonic Committee, who wanted Barnes Wallis, the Vickers company's assistant chief designer, to join them. Wallis was a man of immense ingenuity. Eric had already met with him in October 1944, when working on the Highball project, another of Wallis' imaginative and groundbreaking (sometimes literally) ideas for weapons. But Wallis, as Eric acknowledged, was a man of many ideas besides weapons of mass destruction. He had designed the successful R100 airship and the geodetic airframe construction, which led to the Wellington bomber, and he had even been brought into Supermarine to advise on the Type 300 (later to be called the Spitfire) when the company did not perform as its parent, Vickers, wanted. Despite all of this, Eric had reservations at this time about Wallis joining the Committee. 'Since the Dambuster Raid of 1943, Barnes Wallis had been emotionally affected by the number of casualties suffered in that operation,' he wrote, 'and had declared (frequently) that he would seek never to send any young airman to his death if he could possibly avoid it. This could do much to explain his attitude to what he considered the high-risk M52 . . .'

'Wallis' philosophy seems to have had its influence on Lockspeiser,' he continued, adding that that included the use of one-third-scale models instead of real aeroplanes. This would have been an anathema to Eric. In any event, this use of models resulted in a Vickers contract for twenty-four rocket-powered models to research the government's E24/43 requirement. It was not a cheap solution either, as, according to Eric, the 'eventual model contract cost £400,000 to £500,000'.

The plan was to use the Messerschmitt Me 163 technologies, which Eric had helped uncover during the Farren Mission, as a fast track. Data on the hydrogen peroxide fuel were, however, still in short supply. The debate was now between using an air-breathing turbojet or a rocket motor. Eric took the middle ground and was interested in a hybrid system. Wallis disagreed and, although it is not recorded, they may well have had some heated discussions at the Supersonic Committee. Evidence for this is to be found in Eric's notebook of the time, where he writes that he found Wallis 'a severe personality, who was unlikely to suffer fools gladly, and who was forceful in his views, and somewhat intolerant of those who questioned those views'. A different character, Eric thought, from the one portrayed by Sir Michael Redgrave in the Michael Anderson film *The Dam Busters*.

For the first flight of the M52, the plan was to use the long runway at Boscombe Down, the same runway which would later be used for another ground-breaking, ultimately doomed world-beating British development, the TSR2. By early 1946, the tests were going so well that Eric believed that they were 94 per cent ready to go. The first flight was even planned for October, when Eric might well become the first man to break the sound barrier. However, two serious problems then hit the programme.

The first was in mid-January 1946, with the resignation of Whittle, the jet engine's inventor. Whittle held most of the patents for jet propulsion, but the Labour government insisted on nationalization in July 1945, and the amalgamation of Whittle's Power Jets company with the gas turbine department of the Royal Aircraft Establishment. It might have seemed simple in Whitehall, but at Farnborough, such were the differences in approach to powering aircraft, it was like combining a rugby and a soccer team.

The second issue came the following month, when the Prime Minister, Clement Attlee, issued a directive to suspend all M52 research and development work that would not soon lead to an operational capability. The sound barrier – flying faster than the speed of sound – might have been the 'holy grail' for many in the aviation community, but with the war coming to an end with Germany, and now Japan, the government in Westminster was now keener on developing a civil aviation industry which would lead the world, i.e. long-range passenger airliners capable of flying non-stop to New York, the Middle East and Cairo, en route for the outstations of the Empire.

There was an immediate scramble to justify continuation of the supersonic flight experiments, but Eric knew that, despite the brilliant work of the engineers at Miles, and the boffins at RAE, the project was doomed. Britain was bankrupt, and the Americans would want 'more than a pound of flesh', as he put it, to bail out the country. In the event, the Ministry ordered the M52 fuselage to be chopped up, in what Eric would always believe was 'deliberate destruction'. He was devastated by this decision. Not only did it mean that he would miss out on making history, but it also marked the death knell for Britain leading the march into the next frontier of supersonic flight.

This all left Eric feeling very bitter. For the rest of his life, he was adamant that the 'glittering prize of supersonic flight' had been snatched from Britain and given to the Americans, with Chuck Yeager credited with being the first man through the sound barrier in the Bell X-1 on 14 October 1947.

It was the sharing of British research with the Americans that particularly rankled with Eric. During the war, the Atlantic Charter saw Britain forward all information and research regarding the prospect of supersonic flight to the US. At Langley Memorial Aeronautical Laboratory in Virginia, a committee was subsequently formed to examine the British information, with Bell Aircraft selected as the US national constructor most likely to build an American version of a sound-barrier-breaking jet.

However, Eric revealed that there were no reciprocal briefings between the American and British teams. It was very much a one-way street. The Americans came to Farnborough, but the British were not invited to Bell. He therefore believed that the Americans took all of the British research for M52 and, using it along with their own, produced the Bell X-1.

The Bell X-1 timelines do not necessarily support Eric's theory. It is just possible that American researchers were working in parallel with those of the Royal Aircraft Establishment. As David Weinel wrote in the 'Transmission' column of *Aerospace* magazine in February 2018, 'There is no evidence that the M52 data blithely handed to the US influenced the Bell X-1 project but the similarities between the two airframes are remarkably and coincidentally similar.'

On 4 February 1998, at a symposium looking at the technical and political developments which led to flight at twice the speed of sound, there was considerable discussion about the M52 project and the US development of the Bell X-1. According to Brian Abraham, an authority on supersonic flight, the first record of an American being invited into the project was in 1945, when Dr Clark

Millikan of the Guggenheim Aeronautical Laboratory at Caltech, on behalf of the US Navy, visited the Miles factory at Reading, where he was hosted by George and Blossom Miles and the Ministry of Aircraft Production M52 representative, Group Captain Alan Bandidt.

The following day, Millikan visited the Royal Aircraft Establishment at Farnborough, where he held discussions on the M52 with Ronald Smelt, who was in charge of flight research. Millikan's subsequent conclusions and recommendations on the M52 were that

> the project of a transonic research airplane [sic] is a valuable and interesting one. In general, the design appears to have little to offer US designers, although the very thin biconvex airfoil section with thick constant curvature covering, the unbalanced power boost control and the all-moving tail might be of interest. It is believed that with the present state of knowledge a much better solution of the transonic research airplane problem is possible.

In light of this, Millikan concluded that the design offered relatively little of interest to United States aircraft designers and recommended that no further steps be taken by the (US) Navy at this time to obtain additional information on the project from the British government.

To what degree the M52 project influenced the Bell

X-1 it is impossible to really say, with all of the principal characters having now passed away. While it seems that the Bell X-1 was not a copy of the M52, the final word on the debate must go to Eric, who wrote in his foreword to *Miles M52*: 'For our proud nation [the cancellation] meant betrayal of our leading position in high-speed flight technology.' He continued:

> I think we had lots of wonderful, sound ideas, but underlying that was the solid fact that the Nation was broke and that we were aiming at something that we could not under any circumstance afford. We had to be realists and . . . we went along with some of these things because we felt if we didn't do one or two of them [projects such as the SR53 and TSR2], we'd drop out of the race altogether.

He also argued that it was politics, and not the inability to be a 'highly innovative aircraft industry', which caused the British aeronautical industry to decline.

Frustratingly, it appears that the British committee was on the right track with the M52. In 2018, Richard Gardner, chairman of the Farnborough Aero Sciences Trust, revealed as much when he wrote to the Royal Aeronautical Society in *Aerospace*:

> A rocket-powered large-scale model of the M52 shape, dropped from a Mosquito nearly two years

after the project cancellation, as part of Operation Neptune[1] (which saw a variety of powered models tested) did reach Mach 1.38 in level flight, validating the practicality of the M52 design to go supersonic.

Despite the scrapping of M52 and missing out on the sound barrier record to Chuck Yeager, Eric remained determined to break the sound barrier himself. In his final few weeks at Farnborough, a unique but dangerous opportunity came his way: testing the tailless de Havilland DH 108 Swallow swept-wing jet. This had been designed after analysing the German Me163 Komet, but rather than a rocket motor it was instead powered by a jet engine. Still, it turned out to be just as dangerous as the Komet.

By the time Eric was assigned to the project, Geoffrey de Havilland Jr had already been killed during high-speed trials over the Thames estuary on 27 September 1946, when the prototype Swallow he was flying disintegrated at a calibrated Mach 0.875 in a dive from 10,000 feet. It was Eric's opinion that de Havilland, at 6 feet 2 inches tall, had hit his head against the cockpit canopy as the Swallow bucked when it approached the speed of sound. Eric declared the Swallow 'graceful but a killer'. He thought the British design at least as dangerous as the Me 163 rocketplane.

1 Not to be confused with the maritime part of the D-Day landings.

The design was modified, and a third prototype successfully flown on 24 July 1947 by the legendary former night fighter ace John ('Cats' Eyes') Cunningham, who was by then the chief test pilot at de Havilland. Cunningham realized that the test programme needed more resources than were available to a single aircraft company like de Havilland and urged that the Royal Aircraft Establishment's Aerodynamics Flight should be brought in to evaluate and test. Eric was the obvious choice for this role, especially as he was a good eight inches shorter than de Havilland Jr. Moreover, Eric not only had the physical stature and attitude of mind to be a test pilot, he also had the experience of having flown the Komet, from which the Swallow's wings and flying surfaces had, quite literally, been copied.

As always, Eric prepared for the test flight with the utmost care and attention, including spending time with Cunningham, the renowned test pilot John Derry[2] and Ron Bishop, the chief designer of de Havilland. Reading matter included the wind tunnel data and previous flight test reports, while he listened to the engineering brief on the newly developed Goblin jet engine power limits. He also gained a good understanding of the phenomenon of control reversal, which Derry later experienced

2 Who gave his name to the spectacular Derry Turn aerobatic manoeuvre, still flown by today's jet display pilots.

as the Swallow approached Mach 1 on a 1948 flight, during which he may well have taken the Swallow beyond Mach 1 and hence broken the sound barrier.

On 6 July 1947, when Eric first flew the Swallow, it was very much just to get used to its low-speed handling and tailless design, rather than any attempt to break the barrier. However, as the trials progressed, Eric became more confident and started to push the Swallow's limits, nearly dying in the process.

Reaching Mach 0.985 during the tests, Eric decided to emulate the conditions within which de Havilland Jr had died. Descending to 4,000 feet, at Mach 0.88, the aeroplane jolted wildly, jerking Eric up and down in his seat, resulting in him smashing his head into the roof of the cockpit. It was only Eric's small stature that saved his life. He knew then that this was how de Havilland had met his end. Thankfully, Eric was able to bring the Swallow under control and lived to fly another day. It had been a truly terrifying experience.

Eric later wrote that three factors contributed to his survival in the Swallow and gave him the advantage over the late Geoffrey de Havilland: the model he flew had strengthened wings; he lowered his seat to give his head better clearance; he learned the lesson about control of the Swallow and did not try to 'chase' the violent movements which might have aggravated the loss of control.

On 15 February 1950, Eric's successor as Officer

Commanding of the Aerodynamics Flight, John Muller-Rowland, lost his life when flying the same Swallow that Eric had recently flown. On 1 May 1950, another experienced Farnborough test pilot was killed flying the first prototype Swallow when it literally fell out of the sky when inverted.

While Britain's, and Eric's, attempts to break the sound barrier at this time were at an end, so was his time as an experimental test pilot. The frantic pace of wartime development had been superseded by austerity, leaving less money for new ideas. On the family side, there were also compelling reasons for him to rethink his career. Lynn was pregnant. Even if money had been made available, Eric's high-risk life as a test pilot was not conducive to a settled and secure family environment. Lynn was therefore keen that her daredevil husband settle into naval life as a regular, rather than reservist officer. It would be less risky, and a regular commission would mean financial security and the opportunity for advancement, perhaps even, one day, to become an Admiral.

Eric understood all of this but it still hurt to leave his life as a test pilot behind. He thrived on the adrenaline it offered, yet even he recognized that much of the excitement and opportunities he had enjoyed during the war, and in the years since, were coming to an end. It was time for a new chapter, and now was the perfect time to depart. As you might expect, though, Eric did not intend to leave Farnborough with a whimper.

With his final day at Farnborough scheduled for 12 August 1949, Eric was determined to fly the revolutionary, but ultimately anachronistic, jet-powered flying boat fighter the Saro SR/A1. There was still a hankering after revolutionary ideas in the British aircraft industry, and the Saro flying boat had been inspired by the need five years earlier to develop a fighter capable of 'island hopping' in the Pacific War. Harnessing a jet engine to a flying boat was fraught with dangers, including the risk of the air-hungry jet engine swallowing sea water. Both the Soviet Union and the Americans were developing their own versions of a naval flying-boat jet fighter, but ultimately all three designs were overtaken by the developments in new, larger aircraft carriers. These warships, it must be said, were to be full of British inventions from Farnborough, which Eric had helped pioneer.

In what was clearly an unauthorized flight, Eric took the Saro fighter to Mach 0.82 in a dive but then hit an obstruction on landing in the Solent, resulting in the flying boat sinking. Eric could have easily drowned without the intervention of Geoffrey Tyson, the chief test pilot, who pulled him from the water. He was hospitalized at Cowes overnight but flew back to Farnborough to pack his bags and collect his logbooks. Eric was lucky to survive and lucky that he was not disciplined for not only an unauthorized test flight but also wrecking a valuable prototype, which, incidentally, still lies on the sea bottom off Cowes in the Solent.

After being signed off at Farnborough, Eric's confidential report was sent to Fleet Air Arm's headquarters at RNAS Lee-on-Solent for ratification by senior naval officers. Eric had logged 3094.55 hours, with 2013 deck landings, and was rated 'exceptional'. However, on receiving Eric's report, Commodore Robert Sherbrooke, the Flag Officer Naval Air Command (and a Victoria Cross winner from the Battle of the Barents Sea), commented, 'I have no knowledge of this officer.' Such indifference was a sign of things to come. While Eric was exalted and respected at Farnborough as a test pilot, in the Royal Navy he was virtually a nobody. For a man like Eric, this was impossible to accept.

PART 4

Full Career 1949–70

After eight years of hazardous test flying, Lynn wanted Eric to settle into a family life, and the arrival of Glenn was the catalyst for a change. Eric was destined for a full career in the Royal Navy, which meant giving up test flying for 'fleet time' and becoming a 'proper' naval officer. Initially, it wasn't the happiest time in his career.

20

Joining the Fleet

On 1 March 1949, Lynn gave birth to a baby boy whom they called Glenn after Glenn Miller, one of Eric's idols. The months that followed should have been the happiest period of Eric's life. He was now a father and truly had a family of his own. Yet it was all clouded with anger and frustration. With his time at Farnborough at an end, Eric felt lost and unsure of himself in this postwar world. He was still just twenty-nine years of age and had his whole career in front of him, but after so much excitement it was almost as if this was now the hangover.

After leaving Farnborough, on 29 August, Eric attended a three-month training stint at RNAS St Merryn in Cornwall. It was the holiday season, and there was no housing available, so the Brown family lived in a rented caravan by the sea. It was a chance to enjoy the sunshine and to finally be able to spend some time together. Despite this, Eric still cut a morose figure: distant, and short and sharp when he did speak. It was clear to Lynn that her husband was going to take some time to get used to his change in circumstance. There was too much to process, and Eric was not a man who could speak openly about his feelings, preferring instead to internalize things until he was ready to blow. Simmering under the surface, he was like a pressure cooker.

In November, the family moved into married quarters at RNAS Culdrose, with Eric set to begin his time in the Royal Navy. If he didn't feel particularly enthused about this, there were at least some things working in his favour. The Royal Navy in the 1950s, still with National Service, would no longer be the class-ridden institution it had been in the 1930s. The wartime emergency of bringing in all comers and the selection of the best officers and ratings to stay for a career had paid off without losing the tradition, efficiency and reputation of the 400-year-old service. This seemed like it would suit Eric, who had always felt out of place, and not from the required background. It also helped that his next role had an element of familiarity to it.

Eric was due to return to 802 Naval Air Squadron, his old *Audacity* squadron, where he had first made his name. There, he was to commence work as a Senior Pilot and Acting Lieutenant-Commander under Commander (Air) Donald Gibson, his second CO on *Audacity*. A key part of Eric's role was to oversee the squadron and to train the pilots for all operations accordingly, while attached to the light fleet carrier HMS *Vengeance*. In doing so, he would be able to use his vast experience to mould the pilots of the future. It all should have been perfect. However, it was a far steeper learning curve than he expected.

Leaving Farnborough behind, as well as the life of a test pilot, was a continual sore point. 'It had been my life,' he recalled, 'and I had lost it all.' Moreover, although 802 was his old squadron, Eric struggled to understand his young pilots' mindset in a peacetime flying environment. After all, this was something of which Eric had no experience. Indeed, many of the pilots were only in the squadron due to National Service and had no overwhelming desire to rule the skies. Eric found he could not relate to this, while his students struggled to relate to him. He also didn't always make things easy for himself.

One of Eric's key duties was the signing of pilots' logbooks and rating their flying abilities. He still believed, however, that as the former Chief Naval Test Pilot, he was the only naval aviator who could be rated 'exceptional' and was wont to write down more junior officers who had achieved 'exceptional' from previous authorizers. This won him no friends and made him unpopular.

Eric's difficulties in adapting to the role are illustrated in a report from Group Captain Alan Wheeler,[1] the Commanding Officer of Experimental Flying at the

1 During these tours in experimental flying he rivalled Eric's total of aircraft types flown with a logged 400 plus (but these did include marks and variants which Eric's figures did not). After his service career,

Royal Aircraft Establishment.[2] On 17 October 1949, he said of Eric:

> This officer has had a very long tour of duty as a test pilot in the Aerodynamics Flight at the RAE and has amassed an exceptional amount of experience in that time. He has also become an exceptional test pilot, with, perhaps naturally, a somewhat dogmatic way of giving his opinions on flying characteristics of aeroplanes. His handling of other pilots under his command has lacked tact at times. He has adopted a somewhat independent attitude towards authority. I believe that he will become a very useful officer and pilot after a tour in a Naval Unit again under service discipline. He has shown very great latent ability.

Wheeler rated Eric in the following way, all out of a possible 9: professional ability (with the inked-in qualifier of 'as a test pilot') 9; personal qualities 6; leadership 4 (not good for a potential career officer); intellectual ability 8; administrative ability 5; flying ability 9 (again with the qualifier 'as a pilot').

In Eric's next report of the period, from November

Wheeler was also the air advisor for the Hollywood blockbusters *The Blue Max* and *Those Magnificent Men in Their Flying Machines*.

2 Wheeler knew Eric well as commandant of the A&AEE Boscombe Down and the Airborne Forces Experimental Establishment at Beaulieu.

1949 to April 1950, his personal score slumped dramatically, with all the criteria for promotion reduced, showing how difficult he found it to adapt to squadron life at sea, with his colleagues of the time calling him 'prickly'.

Rear-Admiral Sir Robert Woodard, a naval aviator who served under Eric, later expressed doubts about the wisdom of bringing him back into the mainstream. He commented simply that: 'Eric should have been allowed to keep test flying; that's what he was the master of, and that's where the navy would have gained most benefit from his skills.'

Indeed, even seventy years later, veterans of 15 Carrier Air Group still recall Eric's inability to be a team player. Lieutenant-Commander David (Shorty) Hamilton remains stark, even caustic, in his comments: 'He really didn't fit into the squadron or with junior officers in the wardroom. There is no disputing his talents as a test pilot, but he found coming under command difficult and being in a front-line squadron very different from test flying.'

An early example of not understanding the mood was seen during a Sunday stop-off in Gibraltar. To Eric, it was normal to fly every day, but the Carrier Air Group was less inclined. Such was the opposition that an enterprising young officer found a local by-law about Sabbath Day observance. Instead of flying, as Eric had ordered, the young pilots set off to the beach for the day. Eric just couldn't fathom why his pilots didn't want to fly every day. Conversely, his pilots couldn't understand why he did.

Nevertheless, when the pilots were inclined to fly, they found that Eric had strict principles. He believed that if his pilots could fly in close formation, they could fly in anything, anywhere and at any time. He drove them hard and could be merciless, expecting his pilots to reach the high standards he set himself. Many failed to reach such a high bar, and he failed to understand why they couldn't do things that he found so easy. It also didn't help that Eric knew just how good he was and couldn't hide his arrogance. Lieutenant-Commander Hamilton was leaving 802 when Eric arrived and recalled the following example to illustrate the difficulties that Eric faced:

> We launched a six-ship formation of Sea Furies to practise formation flying and combat manoeuvres from RNAS *Culdrose*. Winkle put us into an echelon starboard (the fighters diagonally across the sky to the right of Eric as the lead aircraft) and then he turned into and below the formation. We had never practised this manoeuvre and as a result there were Sea Furies all over the sky. It was too much for the squadron, and the five more junior pilots just turned for base, ignoring his calls.

By 25 May 1950, when Eric's Captain, John Cuthbert, reported on him for his Annual Report, it seems that there were still issues: 'A difficult officer to report upon. His high reputation as a test pilot is well known and deserved, but his ability as a naval officer is low in comparison. He needs more sea experience both in carriers and other HM Ships before he can properly fulfil the tasks which his seniority requires.'

While his leadership marks remained low, Eric now only rated 7 out of 9 for his flying abilities. This must have really hurt his ego, which was very sensitive in those days. Some improvement was, however, noticeable, as Cuthbert remarked, 'Popular in the wardroom mess, he made a great effort to fit into the ship's life.'

Eric was recommended for command, but not yet ready to take a squadron to sea, so Cuthbert reasoned he should continue to gain experience in his present employment as Senior Pilot. In another simple but terse comment, Rear-Admiral Charles Lambe, the outgoing Flag Officer Flying Training, and soon to be afloat as the Commander of the Home Fleet's Third Aircraft Carrier Squadron, simply wrote 'Concur'. There was not a lot of love in the mainstream Royal Navy for aviators – the *wafu*,[3] as they were known.

In order to improve the squadron's flying ability, Eric became particularly determined to create a more

3 The term has several possible derivations, including 'weapons and fuel users' or 'wet and flaming useless'.

ferocious, competitive spirit among his men. He therefore organized a series of rugby fixtures, which included a game against the University of Sardinia while in Italy. Midshipman (Air Engineering Officer) Harold Lewis, naturally known as 'Taff' because of his nationality and rugby prowess, recalls how things went badly wrong:

> The Italians played on red clay, like the French tennis courts, and the result was 'gravel rash' for many. Eric took it as a sign of getting stuck in if teammates had to see the medics to have the gravel removed, but his teammates weren't quite so convinced, as the game was so fierce, it became known as the 'Battle of Cagliari'.

It was again a sign that while Eric would push himself to the limits, he couldn't persuade others to want to follow him.

By the next reporting period, November 1950 to May 1951, Eric had been confirmed in his rank and had at least started to redress the reporting balance, bringing up his scores to more than 7 out of 9 on average, including regaining that 9 for flying ability. Captain Ian Sarel reported that, as Senior Pilot of 802, Eric 'is efficient and is thorough and meticulous in all he does'. A note appended to his confidential report read 'exceptionally keen on flying'. This was saying something. It was all Eric was really interested in.

While Eric, with Lynn's influence, had set his sights on a career climbing the ranks of the Royal Navy, it was clear that this was not where his talents lay. The Admiralty agreed, at least for now, especially as it had just the opportunity for a man like Eric. It was one which was beyond his wildest dreams.

Leaving Farnborough was not Eric's last taste of test flying. When the Americans asked for Britain's best naval test pilot to join the team at Patuxent River, there was only one man who could fill the post. Eric thrived in the can-do America of the 1950s and took full advantage of the posting, including regular 'recreational' flights in the Grumman Bearcat, often with the soon-to-be astronaut Alan Shephard as his wingman.

21

America

In the early 1950s, America and Britain were total contrasts. Still riding high from victory in the war, America's economy was booming. New interstates, suburbs and schools were being built, all supported by futuristic giant cars, and modern appliances such as fridges and televisions. In the early stages of the Cold War, Hollywood was also keen to use movies to emphasize the American Dream, showcasing the best of Americana to a worldwide audience. It was no surprise that most in Britain felt a great deal of envy.

At this time Britain was still struggling to recover from the war. With huge debts to repay, the economy remained stagnant at best, while there was no money to fund expensive infrastructure projects, let alone clear many of the bomb sites that still remained. Decent housing was also in short supply, and cities rife with slums, while even rationing was still in operation. Everything in Britain seemed grey and backwards, as if on a one-way ticket to oblivion.

Despite all of this, when it came to aviation, and

aircraft carriers in particular, the two countries still needed each other. During the Battle of the Atlantic, Britain's development and use of the aircraft carrier was seen as a battle-winner. The US followed suit in the war in the Pacific, where the ability to have mobile airfields to seek and destroy enemy naval forces, especially the Imperial Japanese Navy, was vital. Britain's ingenuity in this field, and America's willingness to invest, saw close cooperation between the two allies and led to innovations in technique, tactics and operations.

While an enormous amount of money was pumped into US Naval Aviation, and the war in the Pacific was won, America still lacked the technology and innovative design for a new generation of jet aircraft carriers. In the postwar world this was vital, especially as it was felt they would be the key conventional deterrent to the Soviet Union's Cold War ambitions in the Atlantic, Pacific and Indian Oceans. And so the wartime 'special relationship' between Britain and America continued. Britain provided the technical brainpower through the two Royal Aircraft Establishments at Bedford and Farnborough, and the Americans had the ambition, and the deep pockets, to make the ideas a reality. Thanks to this, in the summer of 1951, one such development was causing quite a stir; the angled flight deck.

On 7 August 1951, during a conference at Farnborough, the chairman of the meeting, Captain Dennis

Cambell, by then Deputy Chief Naval Representative of the Ministry of Supply, produced some sketches of his ideas to improve the flight deck layout of a new generation of aircraft carriers. After the failure of the 'rubber deck', Cambell had been wrestling with the problems of landing jets on aircraft carriers for some time. His sketches showed a way to create a slanted 'runway' line off the centre of the ship, at an angle to provide a landing area for jets. It became known as the 'angled deck' and would allow the aircraft carrier to launch and park aircraft forward of the island superstructure in complete safety, even in the event of a landing aircraft missing the arrester wire and having to go around again. In the words of naval test pilot Commander David Hamilton, then leading the Service Trials Unit, the angled deck was 'the most revolutionary improvement to carrier operations ever'.

With the Americans intrigued by this new invention, it seemed a perfect opportunity for both countries to continue their 'special relationship'. In exchange for access to information on the angled deck, the US Navy offered the Admiralty a place for a naval test pilot in America. The Admiralty had just the man for the job.

The posting came as a surprise to Eric. There was no warning or hint from the naval staff. A signal simply arrived which ordered him to report to the Naval Air Test Center, at Patuxent River in Maryland, known colloquially as Pax.

The history of the naval air station at Pax went back a decade, to when America entered the Second World War in December 1941. It was realized that the country had no naval aviation research, development and proving establishment, so in October 1942, after rapid development, the airfield at Patuxent was taken over by the naval department. The US Navy doesn't do things by halves, and a completely new town was built at Lexington Park to support it. The Test Pilot School was subsequently established in 1945, with Pax becoming a world-recognized centre of excellence for test flying.

While Eric was ordered to attend Pax, it was an order that he was more than happy to comply with. He had been treading water in the Royal Navy for over two years, and it had not been a happy fit. Now he was to return to something that really got his blood pumping, being a test pilot, and in the United States no less, where many of the most exciting naval and aviation developments were well under way. Moreover, away from the austerity which still blighted postwar Britain, America was the land of opportunity, and it also meant an enhanced rate of pay. Yet this wasn't just an exciting opportunity for Eric. With Pax situated close to Washington DC, and on the Atlantic Ocean coast, it was also a dream posting for his young family.

After finally taking his 'End of War' leave in August 1951, which, as extraordinary as it might seem, he had rolled over for six years, Eric was rested and ready to depart for some of the most intense and exciting flying of his career. He would also be carrying with him a remit from the Admiralty, to hand a sketch of the new angled deck scheme to his American boss, Captain E. R. Dixon, USN, head of the Flight Test Division.

On 8 September 1951, the Brown family departed from Southampton on the new Cunard liner RMS *Caronia* and headed to New York. After a rough sea crossing, they continued onwards by rail to New Jersey and were met by Lieutenant-Colonel Harry Abbott, a US Army chaplain at Fort Kilmer. Abbott was a familiar face. He had known Eric and Lynn in Northern Ireland a decade earlier, when he had especially enjoyed listening to Lynn entertain American troops, with her wartime radio concerts on the BBC Home Service. Before the Browns left New Jersey, Abbott managed to persuade her to sing for the congregation at the Post Chapel, where Abbott officiated. While Lynn impressed with her vocal talents, Eric was immediately envious of Abbott's Buick Roadmaster convertible. He subsequently vowed to buy an American car as soon as possible.

After being issued with a driver's licence, Eric purchased a 1946 Buick Special four-door sedan 4-litre motor car, perfect for the then open highways of America. In this, the family drove to Maryland, looking out in awe as they passed huge billboards advertising household appliances, saw the flashing neon signs of roadside diners and marvelled at the tail-finned Cadillacs that roared past them on the way. It was unlike anything they had seen in Britain. The country thrummed with hope and ambition. And this was just a taster.

On finally arriving at Pax, Eric was blown away. 'I was astonished at the size of the place; runway lengths and all that,' he remembered. 'In a country that was criss-crossed with federal airways, the free space that had been allocated in that area for their work I found quite astonishing.' Such was its size that it was a 14 mile drive around the perimeter of the airfield. In 1952, the main runway was even extended to 5,000 yards, with an adjustable mobile arrester gear system on another runway, to allow even the heaviest jets to be accommodated for dummy deck landings and similar trials. It was a place beyond Eric's wildest dreams.

However, before he could get too carried away, his primary duty was to assist with the angled deck. As ordered, he duly handed over the Admiralty's sketches to Captain Dixon. They caused such an immediate stir that a prompt discussion with other NATC personnel was ordered that day. Within hours, the sketch, and Eric's understanding of the concept, was on its way, by courier, to the Bureau of Aeronautics in Washington DC.

Following this, Eric showed the NATC pilots and officials at Pax the general idea of the angled deck, landing a

Douglas Skyknight on a dummy angled flight deck which had been painted on the runway. Such was the excitement this created, it was decided that more dummy trials would be held at sea. The USS *Wasp* and USS *Midway* were duly marked with a 6 feet-wide white line, at an 8 degree angle, from the after end of the flight deck. A further innovation developed during the trials was a 5 feet-wide yellow line, painted 340 feet from the flight deck's after end, called the 'roundown', to indicate where a pilot should apply full throttle to 'bolt' and go around again if he had not taken an arrester wire. With the carriers prepared for testing, from March to May 1952, Eric was one of a dozen naval test pilots to take part in a series of trials.

Excited by these developments, the Navy Department in Washington moved faster than the Admiralty in Britain, and in September 1952, it converted the flight deck of USS *Antietam* to the first true angled deck, with an 8 degrees alignment. In November 1952, Eric subsequently took a team from Pax to USS *Coral Sea* for a series of approaches and touch-and-gos, on which Grumman Panthers and North American Savage bombers were used. When the carrier was brought to British waters in 1953, the STU carried out trials aboard her and proved the constant approach technique for themselves. The result of the visit was that HMS *Centaur* was modified with an angled deck in May 1954, and her sister ship, HMS *Albion*, was also modified a year later.

American pilots, and later British, Australian, Canadian, Brazilian, Dutch, Argentine, Indian and French naval aviators, as well as now the Russians and Chinese – all called *tailhookers* in the jargon – were delighted with the simple modification to their aircraft carriers, which would save many lives and many airframes. Eric's work at Farnborough and Patuxent River was formative in that process of risk reduction, and he used his experience of so many deck landings to better inform the boffins at Farnborough, who led the world in this work.

While Eric helped the Americans with the angled deck, he also found that Pax had much to offer him in return. On his very first flight in an F-86 Sabre jet, a design that drew on Me 262 technology, he finally broke the sound barrier. It was something he had been trying to achieve in Britain for years, and yet here he was, barely into his first few days at Pax, and he had already crossed a life ambition off his list. But as usual, it also got him into some trouble. Some of his fellow test pilots dared Eric to 'boom' the Admiral's house. While the house was safe, the sound of the jet breaking the barrier was so loud it smashed his greenhouse. Eric was fined US$20 and told not to do it again. In spite of this, there was plenty more excitement and close shaves to come.

What Eric had not appreciated – nor had the Royal Navy – was that the Navy Department in America was at least a decade ahead of the British in the development of small, tactical nuclear weapons, which could be carried aboard aircraft carriers and delivered by jet strike aircraft. However, the first weapons were large, and heavy at 4.5 tons, which meant having a delivery aircraft which would be the heaviest flown from a carrier at the time. On 13 August 1945, within days of the first atomic bombs being dropped on Japan, the US Navy commenced a design competition for a carrier-borne naval bomber. This was eventually won by North American Aviation, with its design for the AJ-1 Savage.

On 3 July 1948, the AJ-1 Savage made its maiden flight and carried out trials in the summer of 1950. By the time Eric reached Pax River, the Savage had been adapted to fly the Mk 4, a nuclear rather than atomic weapon, which was developed from the 'Fat Man' dropped on Nagasaki. It was big and cumbersome, but the Pentagon liked it because it was 'GI-proof' and could not be accidentally detonated. This also made it ideal for the rough treatment which being stored in an aircraft carrier would bring.

On 7 April 1952, Eric took a Savage aboard USS *Coral Sea* for deck trials, although he did not record whether they were nuclear-related. He did record, however, that he flew an atomic weapon for the first time when piloting a Douglas Skyraider at Pax River, but such was the nature of the classified trials, this flight is not distinguishable in his logbook of the period.

Not all of these flights were safe. Soon after Eric arrived at Pax, an aircraft accidentally dropped a small nuclear weapon outside of the airfield. Fortunately, it didn't explode. The area was still sealed off for 75 miles in all directions, and the bomb was only discovered after a farmer had found 'this piece of junk', loaded it on to the back of his cart, driving it over rutted roads as he did so, before the stunned National Guard came across him.

The excitement didn't end at Pax. Eric also had the opportunity of flying at the famous Edwards Air Force Base in California. This meant a cross-country trip from Maryland to California, experiencing the delights of America from above. The plan was for Eric to see and understand some of the high-speed work being carried out at Edwards, and he was also eager to meet his rival, Chuck Yeager, the US Air Force test pilot who was credited with breaking the sound barrier in 1947. Eric always wanted to be that man, and it is safe to say there was no love lost between them.[1] Yeager was 'a strange guy', Eric later revealed. 'I didn't find him very academic. His knowledge of aerodynamics of high-speed flight was, I thought, slightly limited. Maybe he just didn't care to talk about it. We didn't ever have what I'd call an educated conversation, but a very good pilot; no argument about

1 Yeager continued the 'spat' even after Eric's death in 2016, taking to social media to enquire what all the interest was in Eric.

that at all.' Perhaps Eric's bitterness at Yeager beating him to the sound barrier record somewhat clouded his judgement.

Part of the Edwards Air Base experience was Happy Bottom Riding Club, a bootleg bar in the middle of nowhere and the favourite retreat for test pilots and Hollywood stars. It was owned by 'Pancho' Barnes, who had been an air racer and barnstormer in her youth. All the pilots loved her. The fact she gave a steak dinner to any pilot who broke the sound barrier during a day's testing was almost certainly a factor in this, although Eric thought her 'a very ugly lady'. It was a very macho environment and allowed the pilots to blow off some steam.

Among the regulars were test pilots Bill Bridgeman, who achieved a world speed record at Mach 1.88 in June 1951, as well as Yeager and Scott Crossfield. Crossfield, a former flying instructor and fighter pilot, had always impressed Eric, particularly after he heard that he had investigated an in-flight problem by removing the instructor's access door on his training aeroplane; he identified the issue and corrected it, all while in a spin on his first solo. He was a man after Eric's own heart.[2]

No American had flown the German Me 163B before, so when Crossfield took over the Dryden Test

2 Eric developed a special rapport with Crossfield, describing him as 'the best pilot I ever met over there [in America]'.

Center rocket work, he sought out Eric. Eric was happy to share all he had learned during the Farren Mission, as well as from flying the Komet himself. Soon after, on 20 November 1953, Crossfield became the first person to fly at Mach 2.0 (twice the speed of sound), while flying a Douglas D-558-II Skyrocket.

The rapport between the men became even closer when Crossfield joined North American Aviation in 1955 and subsequently joined the X-15 programme, testing hypersonic rocket-powered aircraft. During his time in America, at both Pax River and Edwards Air Force Base, Eric had certainly displayed the skills necessary to join the X-15 programme, but there was a significant obstacle standing in his way: Eric's British nationality. This made it impossible for him to even be briefed on the handling of the X-15, let alone enter the cockpit. However, when he was approached to apply for US citizenship and to renounce his British birth and passport, with Scott Crossfield encouraging him to do so, this was a step too far for the patriotic Scot. So the cockpit of the fastest aircraft in history eluded Eric through his own choice; it says much about his character and his sense of honour. It was one of the very few aircraft he regretted never having flown, as by August 1963, Joe Walker had taken the X-15 beyond the biosphere and into space, exceeding 100 km in altitude, and was subsequently awarded astronaut wings. Had Eric joined the programme, perhaps he would have been able to add the coveted pilot's wings of a NASA aviator to his list of achievements and might even have become Britain's first astronaut.

Despite reaching the X-15's highest speed of 4,520 mph (7,274 km/h, 2,021 m/s) on 3 October 1967, an unbeatable record set by William Knight, the project was cancelled the following year. Crossfield said the cancellation of the X-15 research had 'caused grown men to cry'. Still flying in his eighties, Crossfield died in 2006 after a flying accident in bad weather. Eric marked his passing with a note, adding him to his twelve most important figures in high-speed flight, joining the likes of Neil Armstrong and the German designer Kurt Tank.

Back at Pax, Eric was allowed to fly and test almost everything in the airfield. This included the Grumman Cougar swept-wing fighter, which he flew at Mach 1.04, just over the speed of sound, as well as the Panther, Banshee 3 and Skyknight, among many others. Eric also enjoyed experimenting with hydro-skis, which allowed aircraft to operate from a beach or strip of sand.

However, while the array of aircraft was exciting, the dangers were more serious than ever. This was a time when every month American test pilots were dying in flying accidents. In the year and a half Eric was at Pax, five pilots he knew died. The Brown family album records men with whom Eric shared an all-too-brief friendship. Men like Lieutenant-Commander Bert Hendrickson, killed in a Cutlass crash; Lieutenant-Commander Armond Dellallo and Lieutenant-Commander Ted Arnold in a Sikorsky HRS-1; and his close friend Lieutenant-Commander Joe Campbell, who died after crashing a Lockheed T-33. Indeed, Eric came close to joining them on more than one occasion.

It didn't help that as Eric was British, the Americans thought he was somewhat expendable and were happy to let him test the most dangerous projects. Eric was only too willing to play along. When HMS *Perseus* arrived in Philadelphia, boasting the new British steam catapult, which utilized high-pressure steam from the ship's boilers to drive the catapults, the Americans were eager to see how it worked. However, when the day of testing came, it was felt the wind was too strong, and the test with an American pilot was therefore aborted. As everyone became frustrated it was eventually decided that the Americans would meet the Brits halfway. The Brits would risk their pilot, Eric, while the Americans would risk their aircraft. Eric didn't mind, and the tests were a success. However, while Eric had countless lucky escapes in his time at Pax, he was almost a witness to his boss perishing in spectacular circumstances.

Lieutenant Colonel Marion Carl was a highly decorated Pacific War Marine Corps fighter pilot and was now the head of the Carrier Branch of the Naval Air Test

Center. He already had a reputation for being a fearless test pilot in naval aviation and was the first US airman to land a jet on an aircraft carrier – with a little advice from Eric.

On 1 April 1952, Carl elected to complete the testing of the new Grumman AF Guardian anti-submarine aircraft, destined for US carriers, with spinning tests but without the tail-parachute, which had been previously fitted in case of emergency in a spin which could not be broken. Carl asked for Eric to fly the chase-plane and to circle him in the spin until he pulled out. Initial tests had gone well, with Carl and Eric carrying out spins, including inverted ones, without issue. However, while over Chesapeake Bay, things went dangerously wrong.

On entering the test routine, the Guardian flipped into an inverted, flat spin. No matter how hard he tried, Carl found he couldn't break out of it and elected to abandon the aeroplane. This, too, became nearly impossible as the ejection seat failed. Still inside the inverted spin, he was also unable to jump out by parachute because of centripetal force. With the ground getting ever closer, Carl managed to escape out of the blind side at low level, and parachuted to the water below. He was in fact so low when he jumped that, as the aeroplane crashed into the water, he swiftly followed it, with his head popping up through the centre of the splash ring.

Watching all of this occur, Eric called the rescue

flying boat from the Naval Air Station and waited overhead for twenty minutes until it arrived. He was hugely impressed by Carl's cool and detached attitude to the incident, and national newspapers were full of the incident, citing Carl as 'the nation's top test pilot'.

Among the test pilots at Pax, Eric also struck up a friendship with Lieutenant Alan Shepard, who in time would become the first American in space. When Eric met Shepard, he was seen as an outstanding naval aviator but lacking in jet experience. Eric therefore taught Shepard everything he knew about jet operations from an aircraft carrier for his postgraduate work on in-flight refuelling systems, carrier suitability trials of the Banshee, and US Navy trials of the first angled carrier deck. The rapport between them was good, and Eric greatly admired him, although he would later admit that he nearly lost Shepard his commission.

One fine day in the spring of 1952, the two pilots were taking advantage of the single-seat Grumman Bearcat[3] piston-engined fighters on the station by 'carousing' around Chesapeake Bay. Very much led by

3 The Bearcat was a 'crackerjack – the best American piston-engined fighter I ever flew', according to Eric. The g-limits (where aircraft are tested to their physical capabilities) were established at 7.5 positive and 3.7 negative – even Eric blacked-out several times keeping current in aerobatics. It was a welcome relief from the day job, which frequently meant straight-and-level flying trials without deviation.

Eric, they couldn't resist looping the Chesapeake Bay Bridge. Following this, they made low passes over the beach at Ocean City, Maryland, and then, to top it, flew at low level over the base. Eric was fortunate that his tail number was not seen by a state trooper who was stationed on the bridge at the time. Unluckily for Shepard, the trooper did record his number, and he was duly reported.

The following day, both pilots were hauled before the naval air station Commander, Rear-Admiral Alfred Pride, who threatened court-martial and termination of service. Shepard was lucky to have the support of Vice-Admiral John Hyland, a distinguished naval aviator from the Pacific War, who knew and understood Shepard's potential, as well as Commander Robert Elder, a legend from the Battle of the Coral Sea, and Shepard's immediate superior at Pax River. The court-martial threat was subsequently dropped, and in May 1961 Shepard went into space – and history – aboard *Freedom-7*.

Eric came to love everything about America. Not just at Pax, but during his fourteen months in Maryland he travelled widely, thanks in part to gaining a US instrument rating which allowed him to fly on Federal 'airways' to California when required. His family album for this period is also a mixture of travelogues for Washington DC, Mount Vernon and the early colonial settlements in Maryland. On one occasion, after driving 3,500 miles to

California, Lynn wrote in their family album that, 'like Cortez', they 'gazed in awe on the Pacific Ocean for the first time'. The Browns also felt very much at home on the base, where there was a good family atmosphere and plenty of parties to attend. Glenn, in particular, loved to play with the other children. Lynn also found herself in high demand at Patuxent River Officers' Club, where she sometimes sang and made so many friends that Eric had to amend his previously negative attitude to socializing. Looking now at pictures of Lynn's birthday party on 31 December 1951 is like looking into social history, as the men have suits and bow ties, and the ladies are elegant in cocktail dresses and high heels.

Most of all, Eric loved the positive, can-do attitude of America, which was in stark contrast to Britain at the time. Rather than look for excuses not to do something, the Americans removed all obstacles and made the impossible possible. America was also the perfect place for someone like Eric, who was always battling with his identity. Unlike in Britain, there was no class system, and people were judged on merit, rather than on where they had come from or where they had gone to school. For Eric, who thrived on results, and expected success to be met with reward and recognition, this was the perfect environment, particularly at this stage of his life.

However, in November 1952, Eric's time in America, originally posted on a two-year tour, came to an abrupt end. 'It was a Sunday morning and I was on the roster for call-in should something need to be flown,' Eric recalled. 'Sure enough, I got the call and drove into the flight ready to fly. It was a nuclear-release trial, which probably needed to be reflown to retrieve some data, but the data was on the launch constraints of the weapon – which was secret and rated "US Eyes Only".'

Despite this warning, Eric managed to obtain some of the data and passed it, by hand, to the British Embassy in Washington DC, just 50 miles down the road. Whether that information was sent to London in a diplomatic cypher or code, which had been broken by the National Security Agency, is not known. Whatever happened, Eric was summoned the next morning to the Department for the Navy, for an interview with the Chief of Naval Operations, Admiral William Fechteler.

'It was an astonishing meeting,' Eric recalled.

> I was met by his aide and ushered through three rooms, each one getting progressively larger, and the gallant admiral was sitting in this vast room with just the one desk. He said, 'Sit down, young man. I haven't brought you here to discuss aeroplanes; I am going to send you home. You haven't done anything wrong, but your problem is you know too much. I think they've allowed you to fly things I didn't realize they were going to, and we thank you because, tit for

> tat, you brought to us the angled deck and the steam catapult. We're parting company on the most amicable terms, but you know enough.' And that was my departure [interview] over.

In 1952, America was in the grip of spy fever. There were huge sensitivities to nuclear weapons information leaking to foreign eyes, which included the British, America's closest ally. President Truman had only just approved the development of the hydrogen bomb, after the Soviet Union had demonstrated how advanced it had suddenly become by detonating an atomic bomb in August 1949. The Americans believed that the Russians had gained the knowledge for the nuclear test from spying on the West, which was fuelled by the defection of the traitors Guy Burgess and Donald Maclean in 1951 – right at the time Eric and family had arrived in the USA. This coincided with Washington becoming aware of the activities of Klaus Fuchs, a British subject, and David Greenglass, an America traitor, who had collected nuclear secrets for the USSR when working at the Los Alamos nuclear facility, as part of the Rosenberg Ring. The FBI had warned US government institutions that all foreigners were suspect, and that none should be trusted with nuclear secrets. It seems that Eric had fallen foul of this paranoia.

Eric was devastated to be leaving his 'dream job' and

also did not relish uprooting the family from school and the local community in Maryland. Before it all came to an end, there was still time for the family to see a little more of America, after which they took passage aboard RMS *Mauretania* in New York and arrived at Southampton on 28 November 1952.

After so much excitement in America, working alongside some of the finest pilots the world has ever known, in a host of revolutionary aircraft, Eric now faced a return to the Royal Navy. He couldn't be less enthused. Britain was grey and bleak, under a cloud of rationing and depression, while Eric's days as a test pilot were finally over. It was all a recipe for disaster.

Commander Eric Brown was not a natural fit as the commander of a naval fighter squadron in peacetime. His personnel records show that he was a square peg in a round hole, but he did enforce his disciplined style of flying on 804 Naval Air Squadron at RNAS Yeovilton.

22

All at Sea

After thirteen years in the air, the aviator looked to be finally grounded. It was now back to cold, hard reality: if Eric ever wanted to become an Admiral, he needed some sea time in the postwar Royal Navy. He didn't like it, but there wasn't much choice. With this in mind, the Admiralty posted Eric to the Third Training Squadron, based at Londonderry. It certainly wasn't America, but it wasn't all bad as at least it allowed Lynn to live in Belfast and care for her recently bereaved mother.

On 10 February 1953, Eric joined his new ship, HMS *Rocket*, at Rosyth, where he faced six months' duty as a supernumerary in order to gain his watchkeeper's certificate. The ship had just been converted at Devonport into a fast anti-submarine frigate, with training accommodation for young seamen and officers, who, like Eric, needed to complete some additional qualifications. Eric's duties on board included being responsible for the Ship's Office and the Boys' Division (as the young seamen under training were called), as well as being Senior Officer of the Day. This might have been far from America and test flying, but in those early months on HMS *Rocket* he gave it his all. A report dated September 1953, from Captain G. E. Fardell, the Captain (Destroyers) of the squadron, suggested as much: 'He has taken a very active interest in all his executive duties. Loyal, hardworking and intelligent. An extremely capable officer. A good leader with a broad understanding and a keen interest in the welfare of his subordinates – they trust him and they like him.' Fardell thought Eric had great potential and tried his best to help him in any way that he could: 'He has been quick to learn the art of seafaring and given further opportunity at an early date to consolidate this knowledge and gain confidence he is capable of becoming a thoroughly good all-round executive officer worthy of early promotion to the Higher Ranks of the Service.'

In his previous Royal Navy role Eric had been marked as arrogant and aloof. Time, and maturity, may have helped take the edges off, but it is more likely that not being in a flying role was the great leveller. This was where he clearly felt superior to most others and felt he had to be seen to be the best. In contrast, while at sea he had to be far more humble and learn the ropes.

It was certainly an exciting time to be aboard HMS *Rocket*. On 15 June, there was the Coronation Review of the Fleet at Spithead, for which Lynn joined Eric to witness the international gathering of warships. Soon after, Her Majesty Queen Elizabeth used the ship to tour parts of Northern Ireland during her first state visit to the province. As Senior Officer of the Day, Eric took on the duty of Officer of the Royal Guard and met the Queen for the first of what would later turn out to be many meetings throughout the course of their lives. He also met Prince Philip and developed a friendship based on shared naval experience, as well as experiences in pre-war Germany. The Prince, himself a pilot, was also very much interested to discuss Eric's time as a test pilot, particularly in America. This friendship would endure throughout their long lives, and Eric would often allude to some 'runs ashore' with the Prince, some of which, he once described as 'like being back Sub-Lieutenants again'. One can only wonder.

In August 1953, Eric found he was sad to leave HMS *Rocket* but happy to have an opportunity to be back among his fellow aviators, joining the Fleet Aircraft Carrier HMS *Illustrious* at Devonport for some more executive training. The carrier, which had seen extensive wartime service, including the battle for Malta and the Pacific, was now the Fleet Training and Trials Carrier. In September, Eric had the frustrating job of three weeks as Hangar Control Officer, without any official flying duties. After that, ever ready to fly, he did manage to get airborne in the ship's Dragonfly search and rescue helicopter and even performed some rescues.

After a while, the Admiralty, knowing of Eric's

aviation background, found a position they believed might suit him well: command of a Sea Hawk jet fighter squadron based at HMS *Fulmar*, the Royal Naval Air Station at Lossiemouth in Morayshire. However, the role was unusual for an officer in the rank of Commander. Fleet Air Arm squadrons of the time were routinely commanded by a Lieutenant-Commander, a rank lower than that which Eric had now achieved. In time, this would create tensions at Lossiemouth. Moreover, back in a flying role, it didn't take long for Eric to return to bad habits.

Following news of his new posting, the family packed for Scotland and on arrival moved into 'Viewbank', a house overlooking the golf course at Lossiemouth. Wasting little time, Eric decided to stamp his mark on 804 Squadron with a tough regime of formation flying, starting with four, then six, aircraft and finally eight in a formation loop. He was ruthless, and worked his squadron hard, but it met with some success. Eric would later claim that on 1 April 1954 the Sea Hawks of 91 Flight, as it was designated, performed the first formation loop with seven aircraft on the thirty-sixth anniversary of the formation of the Royal Air Force. This was made even more satisfying for Eric as they performed the feat in front of RAF Fighter Command, with which there was always a strong rivalry.

However, Brian Ellis, a young, newly qualified Sea

Hawk pilot with neighbouring 802 Squadron, remembers that Eric's efforts were not always well appreciated: 'There were a number of instances which were humorous at the time but looking back now – being older and wiser – could have been catastrophic. Winkle took his squadron aerobatic team very seriously and regularly organized practices . . . he was so cocky about it that it attracted a negative reaction.'

Despite Eric's pursuit of perfection, even he was not immune to mistakes. Nick Cook, a veteran of those days, and Eric's display wingman, remembers the SSAFA Air Pageant at Yeadon, now Leeds-Bradford International Airport. The SSAFA (Soldiers', Sailors' and Airmen's Families Association – the nation's oldest armed forces charity) events were attended by literally thousands of people, and nowhere more so than in Yorkshire.

On 13 June 1954, 804 Naval Air Squadron's Sea Hawk fighters were scrubbed clean and prepared for aerobatic flying. The team's inner circle, the core of 91 Flight, was Eric, as lead, with Cook as No. 2, the Korean air war veteran Paddy McKeown as No. 3 on the starboard side and in 'the box' (centre back and below) Graham (Shad) Foster. The flight took off from RNAS Lossiemouth and positioned initially to RNAS Streeton and then, because of the weather, flew to RAF Linton, the training base in North Yorkshire, where they spent the night to be ready for the following morning.

The display routine had been thoroughly practised in true Eric fashion, and in the air, he led the same way. However, the cloud that day was heavy, and when Eric called air traffic at Yeadon to confirm he was starting the display there was silence in return. This was not unusual, as air traffic control didn't want to disrupt the pilot's concentration. The four Sea Hawks subsequently went through a dazzling display of precision flying, including the formation loop of which Eric was particularly proud. But there was a problem.

'Looking down, there was a huge crowd, but Paddy (McKeown) and I couldn't see the normal runway pattern one expected from a former RAF station,' remembered Cook. 'In fact, it looked more like a horse racing track.' As the formation completed its display, Eric radioed that it was a large crowd, but the Yeadon air traffic controller said that the team had not appeared over the airfield for the SSAFA event. 'It came as a great shock to Eric, and to us all, that we had displayed over Wetherby Race Course, which was in the middle of a meeting,' Cook said. 'Heaven knows what the spectators thought, but they certainly had a taste of real naval flying!' As Cook chuckled, remembering these events, he added, 'Eric's weak point in flying was navigation.'

By July 1954, the naval reporting system had subtly changed to identify both professionalism and general character, as well as the previous areas including flying ability and navigation. This change did not suit Eric, and his confidential reports suffered as a result. It seems that back among his fellow aviators Eric also could not tame his air of superiority or arrogance regarding his skills and experience. In this case, 804 Squadron had an unusually high standard of aviators, engineers and maintainers, with an excellent aerobatic team, and Eric's manner rubbed many up the wrong way.

In a report written by Commander Bill MacWhirter, Lossiemouth's Commander (Air), he marked Eric's flying assessment as 9 out of 9, while his leadership score was just 4 out of 9. MacWhirter said of this:

> Brown is in the somewhat unique position of being the only Commander in command of a Front-Line Squadron and there is no doubt that it causes a difficult situation viz-à-viz other Squadron Commanders and in other ways. He is a determined and forceful character who drives his officers rather than leads them and although he obtains results, and quite good results too, his could not be called a happy squadron. I would guess that he is at his best dealing with things as opposed to people.

This comment really got to the core of Eric's character. He was a master at handling a machine, but when it came to people, he struggled. Indeed, the conflicts between Eric and MacWhirter were legendary. They were both of the same rank, and there were constant referrals up the chain of command to John Ievers, the air station Captain. It might have been expected that Ievers, as a former naval test pilot himself, would have had some sympathy for Eric, but his assessment of Eric's general conduct was no better than MacWhirter's:

> An officer who is difficult to get to know and who is rather hardbitten. There must be few who feel they can confide in him for he seems to have little warmth to his nature. He tends more towards cold efficiency and hard facts. He certainly has the good of his Service at heart, tries extremely hard and is most thorough in all he does. He has an alert mind and sets about his problems with some imagination.

Brian Ellis uses an example of name tabs on uniforms to sum up what many felt was the main issue with Eric's behaviour:

> When the name tabs arrived, most of us used simply our rank and surname; some used nicknames. Winkle would parade around the Air Station with all his decorations after his rank and name on a larger than usual name tab. It wasn't as if he was the only one with gallantry medals. It didn't really fit the Fleet Air Arm leadership model and perhaps a little more modesty would have endeared him to his fellows.

Even at this late stage in his career, when he had achieved so much, Eric still had to over-compensate. He might have been better served letting his accomplishments speak for themselves, but he desperately needed to be seen, and heard. His inferiority complex was never far from the surface, and he had a chip on his shoulder. Comments such as 'Flying a jet from an aircraft carrier is not for the average pilot' might have been true but certainly didn't make him easy to warm to. Issues were amplified by Eric continuing to mark down very successful and skilled front-line pilots, as he maintained that only test pilots could be marked 'exceptional'.

After less than a year, to great relief for some, Eric's time in command came to an end. It was several decades before a squadron commander would again be appointed at the rank of Commander. In Nick Cook's words, 'It had not been a great experiment for the Fleet Air Arm as the conflicts between persons of similar rank on [flying governance] were too great.'

Things did not, however, improve over the course of Eric's next few postings. At HMS *Goldcrest*, the Royal Naval Air Station at Brawdy, West Wales, he served as Commander (Air), and his reports called him an 'austere and rather dour Scotsman who lives for aviation. He is relentless in his war against inefficiency and aims at perfection in whatever he does.' 'Short of homicide,' one report said, he will 'get rid of anyone who fails to aim at his own tremendous standards . . . He does not make friends easily.'

At Brawdy, Eric first met, or rather confronted, Ray Lygo, who was already making a name for himself as a naval aviator. Lygo had joined as a Naval Seaman during the war (as had Eric) and would progress from Rating Pilot to Admiral during his career, ending as Vice Chief of the Naval Staff in 1975. At Brawdy, Lygo had command of a Sea Hawk squadron with bright-red-painted fighters. Eric took exception to these and to Lygo's leadership style. There was also jealousy as Lygo had also flown jets off US Naval carriers on exchange and taken the naval jet squadron to sea. He was already destined to become the youngest Captain and frigate command. It was almost inevitable that they would clash. Eric gained the upper hand then, but later this bitter relationship did not help Eric's chances of warship command.

In November 1956, Eric was posted to the Joint Services Staff College (JSSC) at Latimer, in Buckinghamshire. Captain T. L. Eddison, the head of the naval division, once more voiced reservations about Eric, as so many others had before him: 'He started the Course with an almost infinite capacity for being a bore. His outlook and service interest seemed to be entirely one track and he could not even ask a simple question without giving a lecture.'

Still, Eric was undeterred. Come what may, and

despite all evidence to the contrary, he believed he would become an Admiral. It was all or nothing. Therefore, in May 1957, the Brown family packed up their temporary home at Pressmore Cottage, loaded their new Vauxhall Victor saloon with their belongings and headed north to RAF Manby in Lincolnshire. Manby was the flying college to which all aspiring British aviation commanders were sent, and Eric, the only naval officer, attended No. 10 Course for nine months. There was at least another accomplishment to mark off in June. Having previously broken the sound barrier countless times while in America, Eric finally did it in a British aircraft, flying a Hunter Mk 4 at Mach 1.05.

In January 1958, with the course over, Eric was posted to RNAS Eglinton, then known as HMS *Gannet*, in Northern Ireland. There he was to learn to fly and captain the Fairey Gannet, an anti-submarine aircraft, with a unique contra-rotating airscrew, driven by twin turbo-prop engines, mounted co-axially. In doing so, Eric was preparing to lead a naval team to help the Germans operate the new aircraft, which the newly formed West German Navy, the Bundesmarine, had ordered to complement its Sea Hawk fighters.

While this was an enjoyable posting for Eric, who always loved being in the air, and working with his German counterparts, it seemed yet another job leading nowhere. However, he was wrong. Once more his almost unparalleled aviation knowledge, coupled with his admiration for Germany, offered a dream opportunity, one that he hadn't even considered.

In March 1958, he was told he was to be head of the British Naval Air Mission to the Federal Republic of Germany. For once, the Admiralty put a round peg in a round hole. It was a job he was uniquely qualified to undertake. He was ecstatic.

Yet postwar Germany was very different from the country he had grown to love before the war; there was still an underground Nazi threat with which to contend, as well as the communist East.

Eric was posted to Germany in 1958, when it was coming out of the post-Nazi purdah and starting to reform its armed forces along NATO lines. Britain supplied the newly created navy with fighters and anti-submarine aircraft, and trained the aircrew to British standards. Eric and Lynn were very much at home in Kiel and worked hard to foster good relations. The fighters supplied were the Armstrong Whitworth-built Sea Hawk when declared operational in August 1958.

23

Return to Germany

Since the war, the West's relationship with Germany had changed dramatically. Rather than keep the country under heavy sanctions, and forcing it to disarm, it was now seen as crucial to allow the West German government to reform its armed forces under direct NATO supervision. The threat from the Soviet Union and the Warsaw Pact nations, such as East Germany, was ever growing, and West Germany now looked to be the first line of defence.

Britain's role in this included creating a naval aviation branch for the new West German military, the Bundeswehr. The new service was to be known as the Marineflieger (naval fliers), and Eric, in heading up the British Naval Air Mission, was the man in charge of setting it up. In effect, Eric was going full circle. From once being in awe of German pilots in pre-war Germany, and being taught by the great Ernst Udet, he was now going to teach the next generation of German naval aviators how to fly.

On a bleak January day in 1958, Eric, Lynn and Glenn settled in for a week by the Baltic Sea, staying at the Meersblick Hotel in Schilksee, a suburb of Kiel, which was close to the Kiel-Holtenau naval air station and headquarters of the new Marineflieger. Eric knew the area well from his time leading the Farren Mission in 1945.

By the time the Browns arrived in Kiel, the British advance party was already settling in. This included Commander R. C. R. Hallett, an engineer officer and former test pilot, together with five naval ratings. Two other specialist officers had also been posted to the mission: Lieutenant T. R. E. Beeby for air engineering and Lieutenant F. T. Woosnam, a specialist in electrical and armament systems. After a period training the new German pilots at RNAS Lossiemouth, Lieutenant Brian Ellis had also arrived.

The Germans, under Ellis' command, were capable aviators, but there were some hurdles to overcome. The older pilots, some of whom had served in the Luftwaffe, had not flown in fifteen years, during which time aviation had made huge strides. While they were somewhat behind the younger recruits in terms of tactics and modern flight, they more than made up for it with their leadership, and in being a steadying influence.

There was also an issue with the aircraft that had been commissioned for the West German squadron. The Sea Hawks and Gannets, which Eric had been working with at RNAS Eglington, both lacked a two-seat trainer version for new pilots. The competence of the trainees therefore had to be deduced from their training records, until such time as the Germans bought enough French-built Fouga Magister tandem-seat trainers for grading, which turned out to be several months after the mission started.

However, while the British naval team that had been assembled was a strong one, they despaired at the facilities that greeted them in West Germany. The main operating base at Schleswig had been a Luftwaffe night-fighter station during the Second World War. Since then, no one had bothered to maintain the infrastructure, such as offices and hangars. 'Schleswig airfield was a real shambles,' remembers Ellis. When he arrived direct from Lossiemouth he recalled that a NOTAM (a warning notice for airmen) warned 'Beware of the steam-roller on the threshold.' They were right. A broken-down steamroller had been left abandoned on the main runway since 1945, and Ellis had to land over it. He even remembered that, as he taxied in for the first time, he saw one of the groundcrew busily splitting logs to create aircraft parking chocks because there were no issued ones. Things were so bad that Eric had to go to the local garage in Schleswig and buy oxygen for the aircraft systems, because nobody had thought to order any. It also didn't help the morale of the German pilots that they had to sleep in one of the hangars, on camp beds, with no heat, in the middle of winter.

'Gradually, the facilities were improved,' Ellis recalls, 'once the Germans realized that it was up to them to provide.' The aircraft industry representatives helped out too, undertaking many tasks well outside their normal terms of reference. This included engineers from Armstrong Whitworth, which built the Sea Hawks, buying messing provisions from their 'excellent' allowance.

The distance between accommodation in Kiel and work in Schleswig meant that the mission members spent Monday to Friday at Schleswig, returning to Kiel and their families for the weekend. As all Royal Navy members of the mission were married, and had quite large families, they lived in married quarters near Kiel-Holtenau, a former seaplane base. The accommodation at the Schleswig mess was comfortably used by a mix of Royal Navy, Royal Air Force and contractors, as well as families, resulting in a real community spirit. The bar was well stocked, and, as Ellis recalls, 'At one end was the wartime Malta Spitfire ace Ginger Lacey turned fighter-director telling "it" as he saw "it" and at the other was Winkle, telling "it" his way. The clever officer would gravitate from one to the other during the evening.'

On the social front – what today would be called defence engagement – the Browns excelled. Lynn's ability to charm, and her clear love of meeting people, was rewarded by the number of German friends the couple amassed. In a way, Lynn compensated for her husband's more limited social skills. 'He was not really a people person,' Ellis recalled.

When they had arrived in Kiel, Eric had joined the Deutscher Marinebund. This umbrella organization for German maritime activity had recently been reformed after abeyance in the Second World War and was a mixture of old comrades, as well as being a lobby for shipping interests. It still exists and is still influential. Lynn was so popular at these events that she was appointed an honorary Lieutenant-Commander, in order that she could take part in more naval events. She was the first woman so honoured. She also took command of the social scene and organized '*wiener* roasts' for the British and German families, creating a spirit of camaraderie and bonding which some of the Germans did not expect. Soft power in action before the term had been invented.

Lynn's expert diplomacy certainly rubbed off on Eric, as he also came to understand that the mission was bigger than just training. He needed to be a real diplomat to get the Germans onside and therefore threw himself into winning 'hearts and minds' far and wide. Just as in America, Eric seemed far more comfortable in his skin when away from England. He didn't need to overcompensate for his past and lost most of the bombast that won him few friends.

Together, Lynn and Eric soon excelled at building contacts, networking, and ensuring that the Germans they met saw the best side of the British armed forces, and the Royal Navy in particular. Lynn was busy with the wives, and Eric courted the husbands. A particular example was the Kieler Woche (Kiel Week regatta) in 1959, where attention was focused on 52-year-old Kapitän-zur-See Walter Gaul, the commander of the newly formed Marineflieger. By this time, the Browns were part of the Bundesmarine furniture and accepted as equals – some feat, less than two decades after the war. Such was their popularity, the Browns were invited back to the regatta every year for the next decade, and Eric even had the regatta emblem as a car badge on successive vehicles.[1] Feeling ever more at home, Eric also ingratiated himself when at Christmas he flew one of the German-marked, British-built Saro Skeeter helicopters to Kiel for a NATO Christmas party; naturally the helicopter was carrying the *Weihnachtsmann* (Father Christmas) in the left-hand seat, with a sack of presents for the children. According to Lynn's note in the family scrapbook, with the Skeeter's limited performance, Eric was more than a little concerned about the weight of the presents in Santa's sack.

1 In 1965, Lynn was given the highest accolade from Prince Philip: she became the first female to be allowed, at Kiel, to sail in *Bloodhound*, his pride and joy of a racing yacht.

However, Eric's efforts sometimes led to misgivings from some of his staff, who worried that he spent too much time away in Bonn and Kiel, rather than taking care of the day-to-day running of the mission. 'He had many irons in too many fires', complained one officer, referring to Eric's other role in furthering Anglo-German relations. Ellis was more lenient and understood the diplomatic nature of Eric's job, while also saying, 'He really wasn't a natural instructor.' This was saying something.

When Eric eventually managed to train one of his pilots in one of the Fouga Magisters, he sat up front, while his pupil was 'under the hood' in the back seat, learning blind instrument approach. All was going well until the pupil brought the aircraft in to land and, without warning, flipped it upside down. By the time Eric grabbed the controls, the cockpit was in real danger of hitting the ground. In a flash, Eric pushed the stick and throttle forward and, with the little jet trainer still inverted, managed to climb away from the runway and back into the sky. It had been a very close escape and certainly didn't persuade Eric to spend more time at the airfield. Thankfully, training the new aircrew was only one part of the mission.

Just as they had done in America, the Browns made the most of their time in Germany. They enjoyed touring trips to Norway, the Netherlands, Amsterdam in particular, and to Karlskrona in Sweden, where Lynn and Eric stayed with friends in a summer house on the Stockholm archipelago. The ferry to Finland was also taken, with the Vauxhall running up the miles in the 'land of woods and lakes'. Berlin was, however, the greatest attraction for Eric, as he hadn't been back to the Big City since his last visit in 1938.

Of course, much had changed since then. Not only had the city been through a war and turned to rubble, it had also been split in two. After Germany's unconditional surrender on 8 May 1945, the Allies officially divided Germany into four military occupation zones: France in the south-west, the United Kingdom in the north-west, the United States in the south, and the Soviet Union in the east. Berlin followed suit, with the city split into East and West, with Soviet communism ruling the East, and Western capitalism the West. It would take until 1961 for the Berlin Wall to be erected, but until then these two ideals, and different governments, lived uneasily with each other.

In February 1960, before the Wall was built, and after post-Berlin Blockade tensions had eased, Eric and Lynn took the opportunity to visit the East. They were forced to take a circuitous route, because much of Berlin and its environs were still littered with rubble. Their visit included the Brandenburg Gate – Lynn's photograph shows that a notice, informing pedestrians that they were leaving the West, was then the only barrier between

West and East. The Reichstag was also being rebuilt at the time, and the Russian War Memorial stood in splendid isolation. For Eric it must have been a strange time. He had once seen the city in its pomp during the 1936 Olympic Games. Now it was not only divided, but such was the pitiful state of the communist East, with dreadful working and living conditions, that over 200,000 people fled to the West in 1960 alone. The Wall would, of course, soon stop this exodus.

On the surface it seemed that much of the enmity of the war had been forgotten. There were, however, still many wartime veterans connected to the German services in 1960, as well as a surprising number of former U-boat commanders, with the coveted Knight's Cross of the Iron Cross for valour in the face of the enemy. Being a recipient of the latter did not necessarily mean that an officer was an ardent Nazi, but as Eric and Lynn Brown found out, it might.

Sometime in 1958, Lynn and Eric were invited to a small club in Kiel, where the membership was exclusively former Kriegsmarine officers above the rank of *Fregattenkapitän* (Commander). There were simple rules, and these included that only German should be spoken, and that due homage should be shown to the wartime U-boat arm. During a very convivial evening, the hosts seemed to forget that Eric and Lynn were British and in an extraordinary display of what today would be termed 'political incorrectness' (and, even then, was illegal) the guests were invited to show respect with the Nazi salute to trophies of the Second World War. These included a life-size portrait of Dönitz, the German Admiral who briefly became head of state after the suicide of Adolf Hitler, as well as parts of British and American merchant and warships sunk by submarines, a Kriegsmarine naval ensign and other Nazi-era memorabilia. Eric was 'utterly appalled'. After abruptly leaving, Eric reported the behaviour, and the club's whereabouts, to the British consul in Kiel. Sometime later the club was closed down, but surprisingly no action was taken against its members, despite the flagrant breach of the German criminal code.

While there was still a lingering Nazi underground, the Soviets proved the main threat. Indeed, at this time, at least two Soviet spy rings were operating in Kiel. The geography of the time meant that East Germany was just a few kilometres away, and the sea border was not well policed until the creation of the Inner German Border, and the Berlin Wall, a year or so later. Thus, it was easy for information about the developments in the Bundesmarine in general, and the Marineflieger specifically, to be picked up from gossip and sent off to the Stasi (Ministry for State Security).

The only cell in Schleswig was directed by a German naval air electrical officer, who was apprehended swiftly, having been tracked by the Gehlen Organization.[2] The second ring was based in Kiel and run by a female civilian dentist. Eric always reckoned that he should have been on the alert when she refused to give him anaesthetic for dental work on his damaged palate and lower jaw, the after-effects of enemy action in 1941 and several subsequent crashes. In due course, the dentist was also picked up by the security forces and the spy cell collapsed.

In the years since his work on the Farren Mission, Eric had also maintained his own watching brief on the new Germany and the key players of the former Third Reich. While in West Germany he attempted to make contact with Grossadmiral Dönitz, then in retirement in the Bavarian village of Aumühle, following his release from Spandau Prison two years earlier. He wanted to ask Dönitz about naval aviation in the Third Reich and had been disappointed not to be selected to debrief him after the war. However, by then discussions with the Admiral chiefly concerned politics and governance of Germany rather than naval matters, and there were also plenty of politicians, intelligence officers and

2 Created by Reinhard Gehlen, who had led the *Abteilung Fremde Heere Ost (Foreign Armies Division East)* on the Eastern Front. In 1956, it became the Bundesnachrichtendienst (BND, Federal Intelligence Service) with strong support from the US Central Intelligence Agency and the British Secret Intelligence Service (sometimes referred to as MI6).

Americans queueing to get around the table (and be photographed) with the last Führer. In Eric's view, despite his conviction for war crimes, Dönitz was not an out-and-out Nazi but a professional seaman who just happened to be 'the last man standing in April 1945'. Like many alive at the time, Eric believed that Dönitz saw taking the mantle of Führer to be his duty, but that he also saw that it was his duty to end the war as soon as possible. Although Eric received a cursory response to his request for a meeting, indicating that Dönitz was no longer interested in historical naval matters, he finally met the former Führer in November 1958, although not in an official capacity.

At this time Eric was at Kiel's Nordfriedhof cemetery, attending the funeral of a former U-boat commander who had won the Knight's Cross, Germany's highest decoration. In doing so Eric represented not the Royal Navy, but the Marineflieger. Dönitz was also present and, after the internment, as he walked to his car, he passed a long line of former and current naval officers, which included Eric in British naval uniform, with its gold buttons, rank insignia and naval double-breasted cut, not dissimilar to that of the new Bundesmarine. Passing Eric, he turned and retraced his steps to ask him his nationality. Eric explained that he was British. Dönitz simply nodded and moved on. It was over in a few seconds, but the meeting left its mark on Eric. As

Dönitz left in his car, the German officers present gave him a rousing three cheers – no salutes, as formal recognition of a 'war criminal' was not allowed in the new Germany.

Eric and Lynn were thoroughly enjoying their time in Germany, Eric even more so when an exciting flying opportunity came his way. When a production pilot at the resurgent German aerospace industry's Flugzeug Union Süd, a consortium of Heinkel and Messerschmitt, was being examined for his alleged Nazi past, Eric was invited to temporarily take his place. He didn't think twice, especially as he was also asked to be the post-production test pilot for the licence-built butterfly-tail, French-designed Fouga Magister jet trainers for the Luftwaffe. Right man, right place again.

Sadly, for Eric, the naval mission was terminated on 30 September 1960, with the German naval squadrons assigned to NATO. It was a triumph for all those involved, but particularly for the Browns, who earned many of the plaudits. In their honour, the West German Navy hosted parties and a gala ball at Kiel-Holtenau, at which Lynn wore a startling red dress. Such was their reluctance to leave Germany, they managed to stretch out their final departure until 1 December, when Lynn recorded that 'We finally left our beloved Kiel to return to England.' But what now for Eric's career? Surely, after such success, promotion was guaranteed.

In a confidential report, Vice-Admiral Sir Douglas Eric ('Deric') Holland-Martin, the Flag Officer Air (Home),[3] wrote that Eric's time in Germany 'has fully and with credit justified his appointment with the German Navy'. On the other hand, the Admiral continued, 'On further acquaintance of Brown, COMNAVNORTCENT [the British naval officer with a NATO command for the Baltic], I have doubts whether Brown possesses the personality (or the background) to exercise authority to the high degree required of a Captain, especially if he were to be given a shore command.' Holland-Martin then went on to write, 'In spite of this I have recommended him for promotion now, as, if his appointments were carefully selected, as is practicable for a General List Captain, he seems to have qualities that would make him a great asset to the Service.'

On 30 June, Caspar John, by now the First Sea Lord, wrote in a personal letter to Eric: 'I was so pleased to see in this morning's list that you have been *provisionally* selected for promotion at the end of the year.' On his return from Germany, on 31 December 1960, Eric was promoted to Captain.

The Admiralty, in the throes of modernizing under Caspar John, and readying itself for the amalgamation of

3 He was a former Captain of HMS *Eagle*, later to be Second Sea Lord, in charge of personnel and postings.

departments into the Ministry of Defence, subsequently posted Eric to Whitehall. He was to serve as Deputy Director of the Gunnery Division (Air), where he would be responsible for the development of air weaponry in the Fleet, something he was passionate about. However, it wouldn't take long for Eric to learn why the Ministry of Defence was called the 'Madhouse'.

Eric's staff appointment to the Admiralty in London was a necessary stepping stone to progressing to a flag rank – an admiral. Part of his duties included hosting the Russian cosmonaut Yuri Gagarin, which turned out to be a pleasant chat between two test pilots and confirmed Eric's belief that he would not have wanted to be a spaceman.

24
The Madhouse

In 1961, the First Sea Lord faced a significant problem. For once, it was not the massing of an enemy fleet, or even an existential threat to one of his 200 warships. This time the threat to the Admiralty came from within Britain itself.

By the dawn of the decade, it was already clear that the Royal Navy was no longer the world's premier naval force. Since the war, the front-line aircraft fleet had been significantly reduced to a suitable number for embarkation on just five fleet carriers and four light fleet carriers. Then came the revelations of the long-awaited government report into the future of defence technology, a Defence White Paper, known as the Sandys Report. It intimated that the days of the manned fighter were coming to an end, and that interception would, in future, be carried out by missiles. It even went as far as to say the future of aircraft carriers was limited, and that only a small number of the best active carriers should be retained, although the report added that a new super-carrier was still needed to carry nuclear-armed strike aircraft. Matters weren't helped when the Royal Air Force also claimed that it could cover all areas of conflict from strategically placed island bases, without the need for carriers. However, when the RAF produced a presentation to the Chiefs of Staff, supposedly making its case, Sir Caspar John pointed out that one of their proposed island bases was actually displaced by 200 miles and, without this sleight of hand, would therefore require a carrier.

With no war apparently on the horizon, the House of Commons was, however, loath to pour money back into the Royal Navy. Quite the opposite, in fact. Despite over a hundred warships having been transferred to the Commonwealth in the decade before, the Treasury still demanded that by 1962, fifty-six warships should be retired, and eleven Admirals should go the same way, by which time a new First Sea Lord was due to be appointed, as Sir Caspar John also retired.

Unsurprisingly, Sir Caspar John vehemently disagreed with all of these proposed cuts. Indeed, he believed that aircraft carriers were far more than a mobile airfield. He claimed not only that investment in British aircraft carrier technology such as the Deck Landing Projector Sight and the Height Indicator, developed at RAE Farnborough, was making naval aviation safer, but there was also the opportunity for export sales for cash-strapped Britain. Rather than accept the cuts, Sir Caspar instead championed purchasing new aircraft carriers, as well as aircraft.

At this time, it wasn't just budget cuts that were seen as a threat to the Royal Navy. Just as Eric was arriving at Whitehall in January, Special Branch uncovered the Portland Spy Ring, which had been spying for the Soviet Union. This was a profound success, much welcomed in the corridors of power. But there was also some alarm. In the previous years, Britain had been developing new airborne anti-submarine systems and building its first nuclear-powered submarine, HMS *Dreadnought.* The possible presence of a spy in the Admiralty Underwater Weapons Establishment caused a major rethink of internal security. As a result of these perceived failures in naval surveillance, Sir Caspar created the Department of Naval Security, while stepping up internal counter-espionage.

So when the newly promoted Captain Eric Brown reported for work, on 12 January 1961, equipped with recently acquired bowler hat, rolled umbrella and grey (rather than the traditional pinstripe) suit and naval tie, he encountered a febrile, paranoid atmosphere. He had fervently hoped that this was to be a new and exciting chapter in his career. Instead, he was stepping into a warzone of a different kind, one for which he was not equipped. In his words, he was 'being sent into a full-blown inter-service battle for survival'.

Appointed as Deputy Director Gunnery Division (Air), Eric had one of the most difficult jobs imaginable. While cuts were being proposed, he was responsible for not only searching for any equipment that could give the Fleet an edge, but also persuading his superiors, and then the government, that it was worth the investment. Yet there was another important element to Eric's job. At the height of the Cold War, he also had to watch developments on the other side of the Iron Curtain and gauge just how effective newly discovered Soviet anti-ship and anti-aircraft systems would be. After all, this was the time of Russia's Sputnik satellite and the first glimmerings of manned space flight. No longer, it seemed, could a large fleet of warships disappear over the horizon, when space offered a means to finding and tracking a fleet large or small.

While this was all serious stuff, Eric at least found the job also offered him some exciting opportunities. For instance, he jumped at the chance to meet one of those new heroes of space exploration, the Soviet cosmonaut Yuri Gagarin. On 12 April 1961, Gagarin had become the first human to journey into space, when he orbited the earth in a Vostok spacecraft. So when the Soviet space traveller visited London that July, he was invited to Eric's office in the Admiralty for an official visit. Also present at the meeting was an official named as Ed Brown,[1] a

1 Eric believed he was a Technical Intelligence official.

Soviet Embassy interpreter and a Russian speaker from the Admiralty.

After a handshake, the two aviators took a seat, and Eric asked Gagarin about his impressions of the flight. Still sizing Eric up, the Russian tentatively replied: 'I saw the beauty of the universe from space, particularly our own world, but the main sensation was noise and acceleration on lift-off, followed by peace and quiet in orbit, with little sensation of speed.'

He also laughably claimed, in this Cold War era of propaganda, that the only thing he saw from low earth orbit was 'the huge size of Mother Russia' and he did not notice the Great Wall of China or the Grand Canyon in America. In the interests of diplomacy, Eric chose not to correct him.

As the two spoke, and soon realized that they were both avid aviators, rather than politicians, they started to warm to each other. A sign of this came when the conversation turned to fear, something both men knew well during their careers. Rather than brush it off and play the ice-cold, fearless Soviet, as propaganda might dictate, Gagarin told Eric,'I must confess to a little trepidation on my space flight . . . on the re-entry and landing.'

Feeling ever more comfortable in Eric's company, Gagarin said he considered himself to be a test pilot rather than a space pilot because 'I feel I have greater control over my fate.' When Eric asked about the workload in the capsule, Gagarin replied that he had less to do in the capsule than on a normal test flight. Eric was delighted with that answer, as it summed up his own views about being an astronaut. It never occurred to Eric to want to be one, as in his opinion the flight was more about the technology than the skills of the pilot.

Before the interview finished, Gagarin wandered around the office, where Eric had assembled a host of aircraft models. Gagarin seemed really interested in examining them, especially the American types which Eric had flown at Pax. Rather than be concerned that Gagarin wanted to discuss them as a 'spy', Eric felt his interest was purely as a pilot, and a test pilot at that.

Soon after this, Gagarin returned to what he loved best, aeroplane test flying. Sadly, he was killed in a flying accident in 1968, when his MiG-15 went into an uncontrolled spin from which he was unable to recover. Such were the dangers faced by these brave men and women.

For now, at least, Eric seemed to be settling into his new role well, even if some of the old criticisms still remained. In September 1961, Captain Lewis lauded Eric's 'genuine burning enthusiasm to do what he can to further the cause of naval aviation'. Yet while Lewis liked Eric's sense of humour, he conceded that Eric 'is by no means everyone's idea of a Naval Officer and he could easily rub people up the wrong way'. Still, Lewis respected Eric's knowledge and judgement. He clearly wanted to keep Eric in the Royal Navy but would also only rate his chances of promotion as 'fair'.

Despite this, the Admiralty certainly had a huge amount of respect for all Eric had achieved and saw plenty of benefits in his employment. That October, his autobiography *Wings on My Sleeve* was published, with the encouragement of the Admiralty, who believed it would be a good recruiting tool.[2] It was indeed a rare privilege for any officer to be afforded the opportunity to publish their story, let alone while he was still a serving officer.

The book met with positive reviews in the *Daily Telegraph*, *Flight International* and the Air League's journal, *Air Pictorial*. However, others pointed out some inaccuracies. While these tended to focus more on dates, and aircraft technicalities, we also know now that there were also some flights of fantasy, especially to do with his father's war service and some of his own early flying antics, not to mention some important omissions. In Eric's defence, this first edition was purely intended as a naval recruiting tool and was by no means meant to be a fully fledged autobiography or character study.

Nevertheless, the book pushed Eric into the public eye, and professional recognition of his achievements followed in July 1963 with his election as a fellow of the

2 The final draft was written by the Admiralty's in-house ghostwriter and prolific naval author Kenneth Poolman.

Royal Aeronautical Society. He was asked to address his old school in Edinburgh on a subject of his choosing. Speaking to an attentive crowd of schoolboys, he attacked the sad 'lack of a sense of purpose' of most young people of the day. The fact he chose this subject, above all others, again indicates just how difficult Eric had found it to lead and inspire new naval recruits, who often seemed to him to be from another planet.

However, after meeting the likes of Gagarin, and receiving professional recognition, Eric was tasked with one of the most important jobs of his career. Despite the navy battling against cuts, he was asked to help to research a new advanced aircraft carrier design for the jet age, then known only as CVA-01.[3] This proposed new carrier was to be able to carry over seventy aircraft, and the designs boasted a parallel deck, revolutionary new night lighting and ever more cutting-edge tech that could last for at least the next twenty years. It was a huge show of faith in Eric that its progress was put into his hands. He might not have been a people person, but when it came to aviation, and aircraft carriers, his opinion was greatly respected.

Just as importantly, Eric was also asked to consider

3 It is possible to argue that the Royal Navy did not receive its first jet-age aircraft carrier until the commissioning of HMS *Queen Elizabeth* in 2017. Eric did not live to see the commissioning but was a VIP guest at Rosyth during her building in 2015.

the future mix of aircraft types for the refitted HM Ships *Ark Royal* and *Eagle*. In doing so, Eric sought to have the Fleet Air Arm equipped, for the first time, with aircraft of the same or better capabilities than the Royal Air Force. In his eyes, that would mean a Mach 2 fighter, a low-level strike aircraft, and an early-warning radar type. After much consideration, Eric's first choice of aircraft was the American Phantom, which had just come into service with the US Navy and had broken a string of speed, climb and altitude records. Alongside it, in the strike role, would be the British Blackburn Buccaneer, an exceptional British design now benefitting from sufficiently powerful engines in its Mk 2 guise.

As a result of this, in February 1962, Eric visited the US Navy and McDonnell Aircraft Corporation factory in St Louis and arranged to bring an early F-4 Phantom twin-engined, twin-seat fighter to Britain for trials. In June, a demonstration naval Phantom was subsequently brought to Farnborough, and Eric had the opportunity of flying it off USS *Forrestal* in the English Channel, making touch-and-go landings on the carrier. It was only fifteen years earlier that he had become the first pilot to land a jet on an aircraft carrier in those very waters. During the tests, Eric flew the Phantom at two and a half times the speed of sound, the fastest he had ever flown. At 57,000 feet it was also the highest he had ever flown. Eric was blown away. The Phantom was the best, and he was determined that Fleet Air Arm should have it.

To get the Phantom meant facing down tough opposition. Eric had to fight off not only the Air Staff, but also the civil servants in HM Treasury, who wanted to invest in a British design, the Hawker P1154. However, Eric felt that this new British design, capable of vertical take-off and landing, was not up to the job required at sea, having been designed primarily for land-based use by the Royal Air Force. Eric could also show that the Phantom could go faster, carry more ordnance at greater range and generally do the job better. In his passion to get his point across, he wasn't always the most diplomatic and even admitted that his criticism of P1154 was 'rather too scathing'.

Still, Eric refused to concede. He knew the Phantom was superior to the P1154 in every way, and such was his insistence, and proof, that by January 1964 the Air Staff itself had also become disillusioned with the P1154 and had clambered aboard his Phantom bandwagon. On this occasion, Eric's refusal to compromise was proved to be justified.

Such was Eric's persistence to get the Phantom, his son Glenn remembers that he cut short a family skiing holiday in Austria just to answer some queries for the Ministry of Aviation back home. Indeed, in order to keep on top of things, Eric was constantly back and forth to America to ensure that the Phantom was modified with more powerful engines to suit smaller British aircraft carriers. Nothing was left to chance, and on 8 March 1964, Eric's big moment finally arrived. Accompanying Julian Amery, Minister of Aviation, he attended the McDonnell factory at St Louis to witness the signing of a memorandum of understanding for the US government to provide two Rolls-Royce Spey-powered YF-4K pre-production aircraft, followed by twenty production aircraft, and then later another batch of thirty-nine Phantom jets. This was to be one of Eric's greatest legacies to the Fleet Air Arm. It was his vision, and he had overseen every step.

Eric's boss and old friend, George Baldwin, certainly gave him great credit for his efforts:

> His technical ability and considerable mathematical expertise have enabled him to do a job which, in many respects, could have been done by few other officers in the Navy: specifically, in the past six months, he has put in a great deal of work on the Phantom project and it is largely through his efforts that the Naval Staff was able to establish a firm case for adopting the aircraft.

Things got even better when Eric returned to Britain and was informed that the Conservative government had also given the go-ahead for the CVA-01 aircraft carrier programme. Eric was delighted. This was the way in which he thought the carrier strike capability of the nation should be directed. He wanted robust, fast aircraft designed for naval use and not modified land types. He also wanted the latest weapons and sensor technology and, most importantly, a large aircraft carrier from which they could operate. Now all of this was coming true. However, in October 1964, the Labour Party came to power in Britain, albeit with a small majority, and things quickly changed.

Britain's new Secretary of State for Defence, Denis Healey, was against aircraft carriers. He believed that Britain should never intervene overseas, and aircraft carriers were the perfect vehicle for such intervention. At a particularly fractious meeting on the future of aircraft carriers, Healey was lukewarm on new aircraft and ships which would enable the Royal Navy to challenge the growing Soviet submarine threat. Exasperated at this, Eric was unable to hold his tongue. 'Are you still a communist?'[4] he shouted at Healey, to the astonishment of all those present. Healey had in fact been a member of the Communist Party when a student at Oxford. It didn't matter that Healey had since rescinded his membership. Eric believed he had

4 Eric was not alone in the Royal Navy in thinking that the cancellation of the new aircraft carrier project, CVA-01, had been politically inspired.

dismissed the Soviet threat without good reason and thought it suspicious. Of course, he also hated an amateur like Healey interfering with such an important subject, on which Eric was an expert.

At this point, Labour didn't hold a sufficient majority in Parliament to push through its proposed defence cuts, but after a second election in 1966, the party won with a landslide, increasing its majority to ninety-eight. With little opposition, it could implement its plans for defence, which included cancelling the CVA-01. Eric was furious, but thankfully the Labour government was unable to cancel the purchase of the Phantom jets, as the contract had already been signed. So, in April 1968, the first three British Phantoms, originally destined for CVA-01, were delivered to RNAS Yeovilton.

Despite his success in delivering the Phantom jets, there was still a dark cloud hanging over Eric's chances of promotion. By now, his former reporting officer, Andrew Lewis, had moved to command the brand-new guided missile destroyer HMS *Kent* and had been replaced by Captain Desmond Vincent-Jones, who knew Eric from RNAS Culdrose. Vincent-Jones clearly liked Eric, as is evident in his confidential report. He described him as a 'dynamic little Scotsman' but cautioned that 'he holds strong opinions which when once conceived, he won't change easily'. Such remarks certainly weren't going to push the needle towards promotion. Yet just as things were starting to look bleak for Eric, another opportunity suddenly presented itself.

In October 1964, the creation of a Tripartite Evaluation Squadron (TES) for the Hawker Kestrel, the forerunner to the legendary Harrier jet, was announced. The TES was to be based at RAF West Raynham, with most of the flying taking place around East Anglia, on abandoned wartime airfields. Test pilots from Britain, the Federal Republic of Germany and the United States of America were all involved, as their countries were interested in purchasing the Kestrel. Eric's experience with naval aircraft was always highly sought after, but for the TES he was deemed especially important due to his close links with Germany's Bundesmarine's air wing, with Germany a potential export customer.

While Eric happily looked over the TES, he did not get the opportunity to fly the Kestrel itself. I can find no record of him doing so, even though he added the type to his list of aircraft flown. Sir Don Spears, who was a scientific officer at the time, responsible for authorization of test flying and knew Eric well, also confirmed that he had no recollection of Eric flying the Kestrel. It would have been contrary to all norms, as the Kestrel required a course for which Eric had no time. Twenty years earlier, Eric could have jumped into a cockpit, revised the speed numbers and been up and away. But not in the postwar test flying regime.

In any event, soon after, a chance to once more be posted to West Germany came Eric's way. West Germany was then reinforcing its navy, especially aircraft, rather than diminishing it. The post also offered a whole new set of opportunities, as well as a chance to relive a few memories and meet some old friends. He didn't need to be asked twice.

Although the spy writer John le Carré had resigned from the Secret intelligence Service by the time Eric reached Bonn in 1964, he was a frequent visitor and would stay with the Browns, the location serving as inspiration for his novel *A Small Town in Germany.*

25

A Small Town in Germany

In 1964 the Admiralty asked Eric if he would act as Britain's naval attaché in Bonn, Germany. He couldn't believe his luck. A military attaché is a key member of any legation, and the published protocol shows that Eric was number four in the pecking order at the British Embassy. It was a huge feather in Eric's cap that he had been appointed.

In the 1960s, Bonn was also the capital of the Federal Republic of Germany, West Germany to most, and was considered the most important European posting for any diplomat. It was then even more important than Paris, or even Moscow, because of the British desire to join the Common Market.

In 1957, after the Treaty of Rome was signed, Belgium, France, Germany, Italy, Luxembourg and the Netherlands moved politically, and commercially, closer together in order to help prevent another damaging conflict. Britain was, however, initially excluded, and the government became particularly keen to foster Anglo-German relations in any way possible. Eric's role was therefore crucial in helping to grease the wheels of diplomacy.

His obvious language proficiency certainly helped this process along, as did his aviation feats. The *Guinness Book of World Records* was now available in German, and with his reputation as a courageous flyer he appealed to pilot and lay-person alike. Thanks to this, he was somewhat of a celebrity in German circles in the diplomatic community of Bad Gödesberg, and this made it far easier to navigate and influence, as was his job.

Although Eric's terms of reference remain classified and unavailable, it is clear that he had a remit to form and maintain long-term friendships with the senior German naval hierarchy. To another man, who had been sunk by a U-boat, and seen his best friends die of exposure in the Atlantic, this might have been difficult, but not Eric. He surprised his German hosts by befriending all-comers, including former Kriegsmarine submarine commanders, now senior officers in the new Bundesmarine. This included Captain Otto Schuhart, who had sunk the British aircraft carrier HMS *Courageous* in 1939, and is seen photographed with Eric, in the family album of the time, at Hindenburg Park War Memorial in Cologne, where they paid their respects to the dead of two wars and two countries. Eric was also friends with Friedrich Guggenberger, who sank HMS *Ark Royal* in 1941. While Guggenberger was a former commander of U-81, he had joined the Bundesmarine in 1956, rising to the rank of *Konteradmiral,* before being appointed Deputy Chief of Staff in the NATO command AFNORTH, during Eric's tenure in Bonn. This forgive-and-forget attitude was typical of Eric's professionalism. Indeed, when the former Chancellor, Konrad Adenauer, died, Eric arranged for two Royal Navy motor torpedo boats, which could easily navigate the Rhine, to take part in the escort – with Eric naturally aboard one of them.

As in his previous time in Germany, parties and receptions were very much part of the diplomatic process. Glenn remembers, 'There was a party or reception every single night, sometimes more than one, and Dad was almost always there – even if he did occasionally forget his *aiguillettes*' – referring to an attaché's mark of distinction, worn over the shoulder to denote his or her appointment. Again, this is most revealing. While in Britain, Eric had been criticized for brazenly wearing his naval name tags, decorations and medals. Yet in Germany, he almost played his important role down, being far more understated and relaxed.

Once more, Lynn was also a huge help. She developed a rapid skill in both *Brauhaus Bönnsch*, the local dialect, and the *Standarddeutsch* more commonly spoken at diplomatic parties, while Eric persisted in speaking the *Hochdeutsch* of his pre-war University of Edinburgh course.

'My father became slightly annoyed when my mother was more readily understood than he,' remembers Glenn. 'She had none of my father's background and reading skills in German but would chat away with the local people on trips as if she were a local.'

In his previous posting to Germany, Eric's dedication and diplomatic skills had created excellent relations with the Bundesmarine and the Luftwaffe, now nearly ten years old in its new incarnation, both with their headquarters in Bonn. Eric continued his honorary membership of the German naval veterans' association, the Deutscher Marinebund, and attended frequent formal and informal events. Lynn was again so popular that her 'commission into the society' from 1960 as a *Fregattenkapitän* (Lieutenant-Commander) was reinstated. She was probably the only non-German female to be accorded such an honour. Breaking the mould was what Lynn liked to do: she never regarded anything as 'male only'.

Eric's boss in Bonn was the Ambassador, Sir Frank Roberts, who has been described as an austere but highly accomplished diplomat. When only a 38-year-old minister at the British Embassy in Moscow, Roberts had negotiated the end to the Berlin Blockade and faced down Stalin. Unsurprisingly, he was highly regarded in Whitehall, as well as with Britain's allies in Bonn. He also held an excellent opinion of Eric, which counted for a lot, especially in Whitehall. Although Roberts and Eric had a good relationship, matters were also helped along thanks to Lynn's friendship with Cella, Lady Roberts, the irascible wife of the British Ambassador and the daughter of Sir Said Shoucair Pasha, financial advisor to the Sudan government.

The Roberts, however, had no children, and despite living in Villa Castella, a huge Rhine-side official residence, disliked having children in the Embassy set-up. 'We don't like children in the Diplomatic Service,' Lady Roberts is said to have told the hapless, newly arrived wife of a lowly Second Secretary, whose allocated house was far too small for a family with seven children and another on the way. Housing in the diplomatic world was allocated according to rank, not need, so in contrast, Eric and Lynn, with Glenn at boarding school, had a thirty-six-room villa in a quiet cul-de-sac where other diplomatic families lived. This included the defence attaché of Canada, with whose daughter Glenn struck up a very close relationship on his holiday visits. In fact, the Bonn diplomatic community children bonded well, and connections continued after they had left Germany. Glenn even went on to marry Antonia, the daughter of Embassy official John McKibbin, an Ulsterman who had attended Campbell College, Glenn's old school in Belfast.

Indeed, when Glenn was off school, he fondly recalls

sampling the delights of postwar Germany, where he was almost legally able to visit the *Bierkellers* of Bonn from the upmarket Bad Gödesberg. The area has been described by novelist Hari Kunzru as like a Surrey suburb, with tree-lined streets and large houses for diplomats. Though it was frowned upon by Her Majesty's Ambassador, the older children of the Embassy staff, such as Glenn, occasionally ventured over the Rhine by ferry to the eastern suburb of Königswinter and its Petersburg Mountain sights. 'It had an appealing reputation of being "bohemian" and freer than Bonn,' recalls Glenn. Strangely, Eric never told Glenn that he had stayed in Königswinter less than thirty years before, when he had been travelling with his father, Robert.

In Bonn, Glenn struck up a lifelong friendship with David Franks, the son of British diplomat Arthur 'Dickie' Franks, who was the head of the Secret Intelligence Service station at the British Embassy. Dickie had been posted to Germany in 1962, and he and the Franks family were frequent visitors to the Brown household in Bad Gödesberg, from their official residence in Oderwinter, 10 miles upstream.

In 1977, Dickie Franks was promoted to be Deputy Chief of the Secret Intelligence Service, and a year later became the Service's head. Known as C, Sir Arthur Franks was the first C to be publicly acknowledged. Glenn remembers him as an affable character who encouraged the boys to embrace German life to the full, even in a sleepy backwater like Bonn. He became famous in the SIS for his sense of humour, which one day included instructing his military driver to wait until the usual East German intelligence (the notorious Stasi) tail had arrived. Franks wound down the window and tapped his watch to remind the Stasi operative that he always left at a certain time and he shouldn't be late. The cat-and-mouse spying in the Cold War sometimes had its lighter moments.

One of Franks' team was David Cornwell, officially the Second Secretary in Chancellery, in charge of Anglo-German relations and promoting Britain's membership of the European Economic Community (EEC). By the time Cornwell had moved into the office next to Eric, he had already written several books and short stories under the pen name of John le Carré. Even after Cornwell resigned from the Secret Intelligence Service in 1964, following the huge and, for the British Embassy in Bonn, somewhat embarrassing, success of *The Spy Who Came in from the Cold* the year before, he was still a frequent visitor to Germany. This included staying at the Brown household, where his longhand writing of manuscripts aroused some curiosity. He was, by then, gathering background for *A Small Town in Germany*, which as its central theme has the hunt for pro-Nazi sympathizers in high places in Bonn. Eric felt Cornwell's work denigrated the diplomatic service in general, and the 'secret world' in particular, but he still called him a friend. As might be expected, there were spies on all sides, especially in Bonn, something Eric was soon to find in the most unexpected place.

One spring, probably in 1965 or 1966, deer decided to invade the garden at the Browns' home in Bad Gödesberg. The small German roe deer were as plentiful then as they are now, and beautiful as they may be, they were also considered a pest because they ate everything green. On finding the plant pots disturbed, Eric saw something curious sticking out of the dirt. Pulling it out, he found a microphone and wire, leading to a small transmitter.

The Embassy security team was immediately called and discovered that the electronic eavesdropping devices were found to be of Soviet manufacture of a type well known to Western intelligence. 'We were rather shocked,' recalls Glenn fifty years later. 'We knew we would be under surveillance, and the standard precautions were always taken – my father was a stickler for following the security advice. Now I come to think of it, I expect the Soviets were listening to conversations outside; we were a little more guarded after that.'

Questions also have to be asked about Lynn's work at this time. After her death in 1998, both Eric and Glenn were stunned to be visited by an official from the Home Office to confirm some pension arrangements which dated back to 1965. The Brown family today works on the supposition that Lynn used her frequent contacts with prominent German government officials to probe their sympathies and loyalties. With her gift for the language and socializing, she might have made the perfect spy. So much so that even Eric was not aware of it.

For both Eric and Lynn, their roles became especially important when it was announced in 1965 that Queen Elizabeth, along with Prince Philip, would make a state visit to the Federal Republic for ten days, between 18 and 28 May, including a visit to the divided city of Berlin. As an attaché and fluent German speaker, Eric was to be at centre stage during the visit, for both the planning and the formal duties.

From the moment the Queen stepped off the Andover transport aircraft at Bonn airport, her visit was a huge success. Ordinary Germans lined the streets of Hamburg, Bonn and Berlin to greet the young British monarch on her first visit to the country. In eleven days, across eighteen towns and cities, she made a major contribution to healing the wounds caused by the Second World War.

Historian Simone Derix[1] said that the Bonn government had feared 'that some Germans in their enthusiasm

1 Quoted in the *Guardian* newspaper in June 2015, when Queen Elizabeth again visited a now reunited Germany.

to see her would let slip a "heil"'. In the end, they just shouted: 'E-li-za-beth!' There is also no doubt that having HMY *Britannia* available on the Rhine was a crucial part of the diplomatic planning. Eric relished the pageantry, especially the night of the official banquet aboard the royal yacht. The Brown family albums include a collection of official photographs from the visit, showing the Queen, Prince Philip, the German state president Heinrich Lübke and his wife, Wilhelmine, with Eric at various times during the day, with smiles all round. Eric is usually pictured with sword and medals in his No. 1 dress uniform, whilst the German guests aboard the royal yacht wear 'white tie' evening dress.

While Eric was front and centre during the royal visit, Lynn worked behind the scenes, often coordinating the wives of the Embassy and the wider Commonwealth diplomatic community, ensuring that the Rhine boat tour and various receptions were supported. Years later, Eric would talk of the state visit as setting the seal on the postwar Anglo-German relationship, even if the German political establishment was still not yet prepared to welcome the United Kingdom into the European fold. Conversation at diplomatic parties indicated that Britain's closeness to America was the factor which weighed most heavily with France and Germany when considering the UK's EEC membership.

In August 1965, following the success of the royal

visit, Britain and Germany decided to commemorate the seventy-fifth anniversary of the return of the Heligoland islands to German control after the treaty in 1890 in which Zanzibar was given to Britain in exchange.[2] In the previous five years, Heligoland's two islands had been a bombing range, as well as being protected as a major European migratory bird habitat, a curiously British combination which is not without its merits for nature conservation. On 10 August, along with other Embassy staff, Eric organized for the German military, especially the newly created Bundesmarine, to take part in a major civic event, where recently retired Chancellor Konrad Adenauer was guest of honour. The day was used by the British to continue the charm offensive for EEC membership, even though *der Alte* (the old man), as Adenauer was known, had sided with France's Charles de Gaulle to stop British accession to the Treaty of Rome.

The German Navy also made the most of the opportunity to show off their new jet fighter, the Lockheed F-104 Starfighter, as well as the old Fairey Gannet anti-submarine aircraft, which Eric had helped bring into service, and new fast *S-boots* of the Bundesmarine. They were also keen to demonstrate to the ever-watchful

2 There had, however, been naval engagements there in the First and Second World Wars, and the British had occupied the islands until 1952, when the population was finally able to return.

Soviets, and their Warsaw Pact allies, that West Germany was playing a full role in the North Atlantic Treaty Organization and was not to be underestimated. However, Eric had some reservations about the sleek new Mach 2 machine:

> It was a bit of a handful to be honest with you. I know the Americans were asking all their pilots who flew it to have 1,500 hours [experience] before they were accepted [on to the programme] but the Germans were flying it with 400 [hours' experience]. They were also flying in European weather; not in the blue skies of Texas. It was also 1,000 kilos heavier than the standard Starfighter [interceptor] because they'd turned it into a general-purpose aeroplane. So, if [the young pilots] had an [in-flight] emergency, they were in trouble. If they had one in bad weather, then they were in deep trouble. It was a hot ship.

Such was his concern that he actually campaigned against the Luftwaffe and Marineflieger adopting the Lockheed jet as its fighter and fighter-bomber because he thought the aeroplane was unsuited to the German weather. More importantly, Eric knew that the 1,000 kg extra weight of equipment and avionics required for German service would make it difficult for young pilots to handle at low level and high speed. How right he was – over 260 German Starfighters, or about 30 per cent of the entire fleet, were lost in European operations, often in bad weather or in the mountains – both elements of flying training which had not been taught in America. Such were the mounting losses that the Starfighter attracted the German epithet of *Witwenmacher* (widow-maker), usually reserved for the experimental aeroplanes of the last days of the Third Reich.

Eric always said that, other than the Komet rocket fighter, he had not flown any standard production aeroplane as demanding as the Starfighter. Eventually, the Luftwaffe Inspector-General, General Werner Panitzki, let his distress at the loss of so many young pilots get the better of his political judgement. After he spoke out against the purely 'political acquisition' on 25 August 1966, the Federal Defence Minister, Kai-Uwe von Hassel, dismissed him, and he was replaced by Generalleutnant Johannes Steinhoff. The new appointment was a fighter ace and former Messerschmitt Me 262 pilot, with whom Eric was at least able to forge a relationship based on common experience. Steinhoff was an excellent organizer, leading the Luftwaffe into the jet era and eventually commanding the NATO air forces in Central Europe. He never forgot that a British surgeon had rebuilt his face, after a serious crash-landing in an Me 262 had led to a catastrophic fire which badly burned him.

On 27 May 1967, there was great sadness in the Bad Gödesberg diplomatic community as Eric was tour-expired. The Browns had been well liked, and Eric had proven that, in the right environment, he could excel in a diplomatic role, especially when aided by Lynn. But it was now time to leave the diplomatic life to once more return to the Royal Navy and assume command of a Royal Naval air station on the Moray Firth in Scotland. He did not know it, but the end of his time in the Service was fast approaching.

Eric was appointed to command RNAS Lossiemouth in 1967. It was to be his last appointment in the Royal Navy, although at the time he hoped a good showing would allow him to command an aircraft carrier or become a naval attaché in Washington. Those who served under him vary in their opinions of his success, and 'their Lordships' made a fateful decision in 1970.

26

The Final Command

Eric was closing on his fiftieth birthday and must have pondered his future in the Royal Navy. If the chances of a command at sea were difficult before, they were now almost non-existent. Financial constraints were causing manning levels to be examined, with a programme for 16,000 redundancies to be made between 1968 and 1973. The Home and Mediterranean Fleets were also to be amalgamated into the Western Fleet, reducing at a stroke the numbers of admirals and staff officers needed. There was still a Far East Fleet, but even that was scheduled to be called home in 1971. Worse still, contracts of service beyond twenty-two years were to be limited.

Things were beginning to look very bleak, so in the circumstances Eric's posting as Captain at RNAS Lossiemouth was grasped with both hands. It would mean Eric had another chance, albeit slim, at a much-coveted Admiral's hat, even if command at sea was retreating as the Fleet reduced in size. At least at this stage he was still in the race.

On 11 September 1967, Eric and the family returned to Scotland and moved into the official residence known as the Old Manse. While HMS *Fulmar* was a busy naval air station, designed to help protect the nation from the enemy, its pilots were most at risk from its benefactor. In 1938, the Laird of Pitgaveny, Army Captain James Brander-Dunbar, had ceded the land to the Air Ministry, yet even now, twenty-nine years on, he still thought of it as his land and felt that the aircraft, and the pilots, were there for his entertainment.

From time to time, when the laird was bored, or in a bad mood, he would take pot-shots at passing aircraft with a 12-bore shotgun. 'The noisy Scimitar was the one which reminded him most of a grouse,' according to one local. He feuded regularly with local councils and when he took exception to someone, which he did occasionally, he would demonstrate his disdain by lifting his kilt in their presence; Brander-Dunbar, being a true Highlander, wore nothing underneath. At the slightest grievance, real or imagined, he was known to demonstrate his lack of underpants.

The laird also oversaw the local hunting and fishing scene, which had been most active under previous captains, including Eric's immediate predecessor. However, Eric was far too busy to entertain the laird with hunting and fishing escapades, which the laird took as a slight. He therefore bided his time to take his revenge.

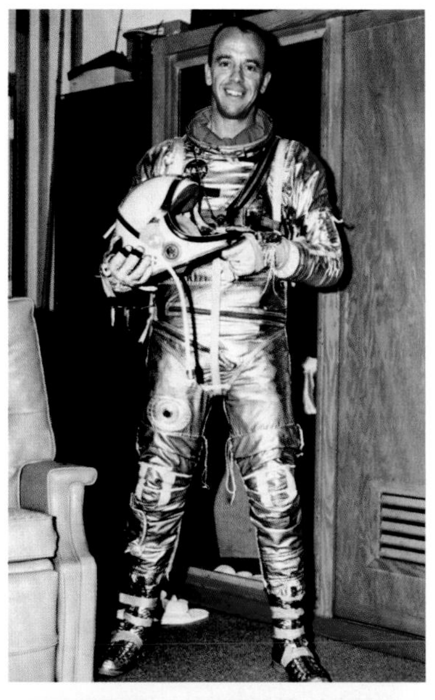

Alan Shepherd was an outstanding naval test pilot with whom Eric struck an immediate friendship. The two would routinely fly a pair of Bearcats around Chesapeake Bay, including a memorable time when they flew under a bridge.

Eric's immediate boss at Patuxent River was Marion Carl, a highly decorated US Marine Corps ace from the Pacific. They shared several hair-raising experiments and kept in touch for years afterwards.

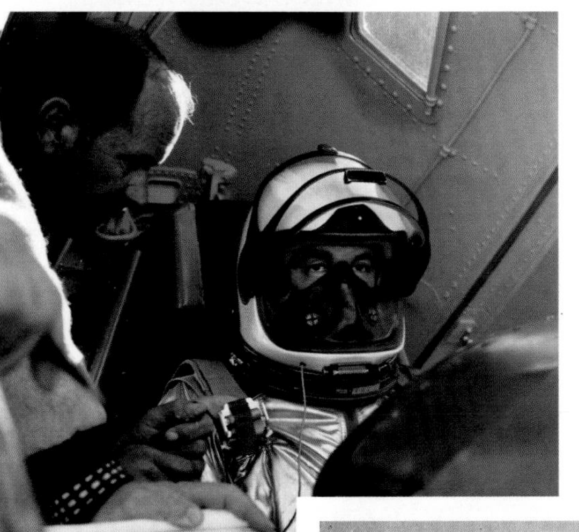

Eric classed Scott Crossfield as the best American test pilot he met at Edwards. Both loved the thrill of speed as exemplified by this picture of Crossfield in a Skyrocket in November 1953. The American was one of Eric's supporters when an X-15 rocketplane test flight was mooted.

For Eric, one of the joys of America was having his family with him. There was also the opportunity of owning a powerful motor car in the shape of a 1946 Buick Special. In it, the family toured the States during leave periods.

Eric joined the frigate HMS *Rocket*, based at Londonderry, on 10 February 1953. It was his first non-flying sea time and he found the change of style disconcerting at first.

The first meeting between Eric and Queen Elizabeth II was in Londonderry aboard HMS *Rocket* in July 1953. Eric is the Officer of the Day, with cutlass.

Eric was not a natural skier, but his wife Lynn certainly enjoyed winter sports, as here in the German Alps during a break from the British Naval Air Mission.

After some carrier time on HMS *Illustrious*, Eric was promoted to Commander and given command of a front-line jet squadron, 804. It was not a particularly happy time as Eric found the legacy of being Chief Naval Test Pilot hard to shake off.

804 Naval Air Squadron at RNAS Lossiemouth formed an aerobatic team with its Sea Hawk jet fighters. The pilots, led by Eric, had a very successful season in 1954. Eric's final report from Lossiemouth rated him exceptional in flying but poor in leadership, remarking that 804 was 'not a happy squadron'.

Eric took over the job of chief production test pilot at Focke-Wulf, which built the Fouga Magister under licence for the Luftwaffe.

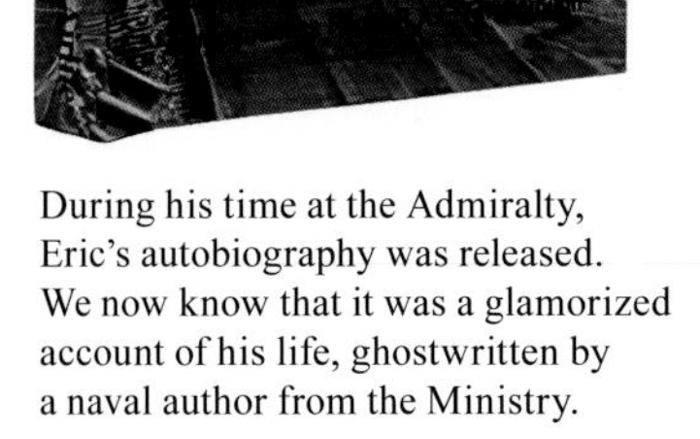

Eric took personal charge of delivery to Germany of the Saro Skeeter from the Westland factory. No surprise then that he flew Santa to a NATO children's party in one.

During his time at the Admiralty, Eric's autobiography was released. We now know that it was a glamorized account of his life, ghostwritten by a naval author from the Ministry.

Eric, as British naval attaché to Bonn, escorted Queen Elizabeth and the Duke of Edinburgh when HMY *Britannia* docked on the Rhine in May 1965.

Eric worked tirelessly to equip the new British aircraft carrier project, CVA-01, with the best available aircraft: the Phantom, Buccaneer and Gannet aeroplanes, and Sea King helicopters.

In his search for the best naval helicopter of its time, Eric also visited the Boeing-Vertol company in Philadelphia to try the Chinook. It was too big for CVA-01 and not selected.

In June 1962, Winkle flew the Phantom for trials aboard USS *Forrestal* in the English Channel. He loved it and persuaded the Ministry of Supply to buy the American fighter, resulting in a contract on 8 March 1964.

By the time Eric commanded RNAS Lossiemouth (HMS *Fulmar*), the writing was on the wall for naval fixed-wing aviation. Nevertheless, Eric was determined to make this his chance to become an admiral.

On 11 March 1970, Eric flew his last sortie as a pilot in command, after over 6,000 hours as a pilot. Fittingly, he choose a helicopter. A test pilot until the last, when the Whirlwind helicopter developed a malfunction, he assured his crewman and passenger that all would be well.

A leading light in the Helicopter Club of Great Britain, Eric helped organize the World Helicopter Championships in Europe, including the Army Aviation Centre at Middle Wallop in Hampshire and at Castle Ashby in Northamptonshire.

Eric forged a lifelong friendship with Sergei Sikorsky, son of the great Russo-American helicopter pioneer.

As President of the Royal Aeronautical Society Eric was frequently invited to events in the US. At Edwards Air Force Base the attendees are pictured in front of the Lockheed SR-71 Blackbird, one of only two aeroplanes that Eric wished he had flown.

In 2010 Eric was the guest of honour at a private gathering of legends in London, including Neil Armstrong; Apollo 13 Commander Jim Lovell; the last man on the moon, Eugene Cernan; and land-speed record holder Andy Green.

WORLD RECORD

GUINNESS BOOK OF RECORDS

Captain Eric Brown
has flown
487 different types
of aircraft
as a command pilot

The *Guinness Book of Records* certificate for the largest number of aircraft types flown. If variants were included, Eric would have flown closer to 750. No one is likely to ever better this record.

On 21 January 2015, Eric was the guest of honour at a birthday dinner at Bucks Club in Mayfair, where he again entertained the guests with stories. He is pictured here with Admiral Sir George Zambellas, the then First Sea Lord.

After his death, a group of admirers in Edinburgh raised a statue to Eric at Edinburgh Airport as a reminder to a new generation about the great pioneering airman who called Scotland his home.

Eric was immensely honoured to be selected as the 3,000th castaway on *Desert Island Discs*, where he charmed Kirsty Young and entertained a whole new generation of admirers.

Proud and confident to the last, Eric will always be regarded as a national treasure and a legend.

Outside the Captain's office was a perfectly manicured lawn, known as 'the Quarterdeck'. The green grass was regularly cut under Eric's watchful eye and woe betide anyone who might stray across the green instead of keeping to the pathways. The result, if spotted by Eric, was a 'right Scottish bollocking' out of the Captain's window.

One fine day – history does not record when exactly – Eric and the laird had a disagreement about something, possibly the ownership of the Quarterdeck. Soon after, Eric's morning heads of department meeting was rudely interrupted by a low mechanical noise. Unable to concentrate, Eric reached the window just as an old grey Ferguson tractor with four-shoe plough, driven by the laird, cut a furrow diagonally across the green swathe. On being sighted, the tractor and its driver beat a not-too-hasty retreat, as the naval police on the gate looked on, bemused. For once, the Captain of RNAS Lossiemouth was left speechless.

Despite this irritation, there were some benefits to Lossiemouth, not least the opportunity for Eric to be airborne every day. As the naval air station was home to training and operational squadrons, it was natural for Eric, as the Captain, to want to fly everything and anything in his charge. However, while all other pilots were expected to abide by the rules when it came to flying aircraft, Eric didn't think they applied to him, as Michael

Layard, then the Senior Pilot of 764 Naval Air Squadron, remembers well:

> Winkle would call me on the telephone and say: 'I have a spare morning, so I thought I would do some flying.' I would respond that there would be a Hunter and one of the QFIs [Qualified Flying Instructors] available whenever he arrived. Noting that, as he hadn't flown for a few weeks, he would need a check ride. The response was always the same: 'I don't need an instructor, I'm a bloody test pilot.' And my answer was similarly robust and simple: 'Sir, they are your rules.' Winkle would go off for some handling and a few trips in the Hunter T8 trainer, which we used to give fighter pilots experience in the world's first *Top Gun* school, as the Americans would call it later. Eric would land-on, and the QFI would step out, and off he would go again. We didn't allow him in the single-seat Hunter fighter,[1] but he didn't seem to mind. We knew he was a perfectly competent pilot and we all admired his flying skills. But he wrote the rules in the first place.

1 Forty ex-Royal Air Force Hunter F Mk 4s had been converted for naval use and called the Hunter GA Mk 11. The modifications included an arrester hook, and some later had a Harley light, which had a lens to diffuse light, allowing for better night landing safety. The guns were removed, but they made excellent fighter-trainers.

This didn't matter to Eric. In his eyes, with his experience, and with the weight of command on his shoulders, the rules should not apply. He felt he needed no additional burdens, such as keeping current, and in a breach of naval rules he would persuade his co-pilots to sign for the aeroplanes he was not 'qualified' to fly.

This can be seen when Robin Spratt, the Lossiemouth Instrument Rating Examiner, was detailed to refresh Eric's instrument rating for flying in all weathers. Eric told him that he had more important things to do than take a course to fly solo, so it then fell to Spratt to fly with Eric in the Sea Prince twin-engined transport or the Sea Vampire T Mk 22 two-seat trainer. Both of these aircraft were used extensively by Eric for what was called 'communications flying', but it almost got Spratt into trouble:

> I had one trip with Winkle to the west coast of Scotland, near Mull, and he landed on the beach [in the Sea Prince]. It was low tide, so the beach was hard, and there was no drama, but I always wondered if there had been, what would have happened (if we had had a mishap) as I had signed for the aircraft, but the pilot flying was the Captain of the air station – to whom I would have had to answer?

Spratt wasn't the only one to be concerned about signing for an aeroplane, and then having the 'unqualified'

Captain of the air station fly it. In 1969, towards the end of his tenure, Eric flew down to the Fleet Air Arm's headquarters at RNAS Yeovilton in a Sea Venom two-seat trainer. The pilot who signed for the aeroplane this time was Lieutenant-Commander Monty Mellor, a highly regarded naval aviator and squadron boss, who 'let' Eric fly back after lunch with his boss, Rear-Admiral David Kirke, the Flag Officer Naval Flying Training.

As they crossed the Scottish border, the weather closed in, with heavy snow and low cloud. With Eric 'unqualified' to fly the plane, Monty became very concerned as to what would happen if there was an incident. As Eric looked to land, Monty said with a straight face: 'Don't make a hash of this, sir. Remember there's fifty years of flying experience in this cockpit.' That relieved the tension, hitting Eric's sense of humour head on. The Sea Venom landed safely, and the station flight was also that year's recipient of the coveted Bambara Flight Safety Trophy for the best all-round safe flying performance by a ship, squadron or sub-unit. Eric was delighted, as this was a real feather in the cap of the Captain and the air station.

Despite this near miss, Eric was undeterred in flying what he wanted, when he wanted, whether qualified or not. After all, in Eric's eyes he had been a test pilot and had flown an array of dangerous aircraft without any courses or manuals beforehand. However, when Eric wanted to fly the new Buccaneer Mk 2, Ted Anson, the Senior Commander, tried to stand firm. Anson told him, 'The modern Royal Navy requires a pilot to be qualified on a new type through the defined training programme.' Eric was taken aback but on this occasion he accepted Anson's word and asked for copies of the appropriate books to be delivered to his office. 'How long is the course?' he asked. 'Six months, sir,' was Anson's reply. 'Humph,' said Eric, 'I'll do it in six weeks.' And, with that, he set out to master the Mk 2's systems in record-breaking time.

While admitting that 'this was quite a complicated machine', Eric locked himself away in his study at the Old Manse almost every night, memorizing the complex series of systems' instructions and diagrams, which made the Buccaneer a bit of a beast but an exceptional strike aircraft. Soon bored and frustrated by this, he decided to enlist the help of Lieutenant-Commander David Howard, the Commanding Officer of 736 Naval Air Squadron, known as the Jet Strike Training Squadron, who was considered 'Mr Buccaneer'. 'Winkle never did a formal course before flying the Buccaneer,' Howard recalls. 'He did two flights with me in the back seat on 10 and 14 May 1968 after I briefed him thoroughly for both. We flew for 1 hour 10 minutes and 1 hour 20 minutes respectively.'

There was nothing wrong with Eric's flying, and

both flights were uneventful, until the second landing on Lossiemouth's main runway, known as 23, by its compass bearing. 'The last part of the runway slopes downhill, and as we neared the last quarter, Winkle squeaked "Brake failure,'' Howard remembers.

> My reaction was automatic – 'Hook down,' I exclaimed, and Winkle, drilled as he was in such matters, dropped the arrester hook, which took the arrester wire which was strung across military runways for just such an eventuality. We stopped before running off the end of the runway, and a tractor had to come and retrieve us. That was embarrassing enough, but at least we did not have the Captain of the air station run off the runway!

However, even though Howard did Eric a favour and was a noted aviator himself, Eric still refused to rate his aviation skills as a '9' in his annual report. Howard was disappointed and believes there was an element of jealous rivalry behind this. As we know by now, Eric believed that he was the only naval pilot who deserved a '9', no matter who his rival was. This was hardly the mark of a natural leader.

Despite his years of experience, it still came as a shock to Eric that the modern Royal Navy valued management and leadership skills more than test-flying skill. This ensured that he remained somewhat unpopular. When he ordered daffodil bulbs to be planted in front of the air traffic building in September 1968, they bloomed in such a pattern that they spelt out a rude word. It has never been confirmed who oversaw the planting, and sadly no photograph has been found. The message was, in any case, only visible from the air. Perhaps it was the junior rating working party, who were not enamoured of Eric, who, on arrival, had doubled the size of the brig (prison) and barred the windows to the rum store; the Royal Navy still had the tot as part of its privileges in those days.

On occasion, Eric could show compassion to the ratings and junior officers. On New Year's Eve 1969, Eric, in full mess dress, decided to leave the wardroom party and took Glenn with him on a tour of inspection. In those days, the air traffic control of naval air stations was maintained 24/7 and 365 days a year; New Year's Eve was no exception, even if no aircraft were flying or expected. On reaching the visual control room, Eric found two watchkeeping ratings asleep. Glenn remembers that 'instead of berating them, my father woke them and asked if they had had dinner. When they replied no, he sent them away to their mess and told them to report back in half an hour; he would stand their watch. Or rather he and I would. It showed me my father in a completely different light.'

Vice-Admiral Donald Gibson, a real mentor to Eric,

was, however, realistic about Eric's promotion chances in the reduced Royal Navy and only rated them as 'fair'. Gibson's successor in post, the Taranto veteran Vice-Admiral Sir Richard Janvrin, said the same nine months later, adding that Eric was 'eminently suitable as a Flag Officer for certain specialized jobs such as Deputy Controller Aircraft (RN) or Director General Aircraft', but the Admiral reiterated that Eric's promotional chances would be fair 'on an absolute basis'.

The promotion issue was simple enough. There were too few jobs for too many qualified applicants. Being famous, and having more test flying experience than any other naval pilot, would not suffice alone. Eric's next confidential half-yearly report in November 1968 said as much. It started off well, describing him as 'a small, fit, grey haired Scot showing little sign of his 49 years, still bursting with enthusiasm for the job in hand, and possessed of a terrier-like tenacity of purpose', but it ends with an interesting summary: 'With the changing shape of the Navy I am not wise enough to assess this unique character's chance of selection for the Flag List.' The Royal Navy was doomed to contract, and it needed all-round leaders; Eric's score of 6/10 for leadership would not have helped, nor the 6/10 for tact and cooperation.

Competition was hard, and all-rounders were needed, so a 'unique character' like Eric did not fit comfortably with the more homogeneous requirements of the Cold War Royal Navy. In the Second World War, someone like Eric would have been welcomed for key tasks and special roles, but the Cold War was a different time, when such officers were seen as expendable.

Eric could sense the writing was on the wall, but he thought, perhaps, that he might again be put forward for a diplomatic role abroad, where he had previously excelled. Indeed, the role of naval attaché in Washington DC, as part of the British Defence Staff, was available, and this would have suited Eric very well – and Lynn would also have loved to return to America. On paper, Eric had all the right qualifications: previous experience with the Diplomatic Service, a pro-American attitude, a close working relationship with the US Defense Department after his time as Deputy Director of Naval Air Warfare, and a personal reputation, which was perhaps stronger in America than in Britain. But he had crossed swords with Denis Healey, and the civilian machinery of the Ministry of Defence, once too often.

Whitehall insiders later told Eric that Healey was adamant that 'that man' would not go to Washington or command at sea and, 'if he had his way', would be retired as soon as possible. Healey also did not want to emphasize naval aviation to the Americans, when he had just ordered the end of fixed-wing flying at sea; neither would he listen to the Second Sea Lord's[2] view that after Lossiemouth, Eric would also be well suited to be Flag Officer Naval Flying Training. The latter was reported widely in the Fleet Air Arm, including a mention in the restricted circulation magazine *Flight Deck*, which kept everyone informed of policy. In short, Eric was finished.

When Eric received the news that he would not have a further career in the Royal Navy – the 'scuttlebutt' as it was known – the feeling of failure, perhaps the first in Eric's life, was all-pervading and self-destructive. It would have been rammed home soon after when the newly appointed Commanding Officer of HMS *Ark Royal*, a post for which Eric had lobbied for himself, arrived at Lossiemouth. Raymond Lygo and Eric had crossed swords at HMS *Goldcrest*, the Royal Naval air station at Brawdy, and now Lygo, four years his junior, was being given the command which Eric had coveted. Relations quickly deteriorated even further.[3]

At this time the *Daily Mail* announced a transatlantic air race between the Post Office Tower in London and the Empire State Building in Manhattan. If the Royal

2 The Admiralty Board member responsible for personnel and postings.
3 Under Lygo's command, HMS *Ark Royal* would cut a Soviet destroyer in half during the Russians' harassment of flying operations off Crete in November 1970. It was the Russians' fault, but Eric was secretly pleased. However, the incident did not stop Lygo becoming the Royal Navy's top sailor a few years later.

Air Force was going to enter the new Harrier 'jump jet' then the Fleet Air Arm also wanted its Phantom and Buccaneer aircraft in the race. Lygo casually mentioned that he was going to fly the Phantom himself, which left Eric furious, especially as he was the man responsible for them being with the Fleet. With Eric relegated to the sidelines, Lygo, and the Phantoms of 892, won the race in a little over four hours. Eric found it hard to keep his envy in check. In his eyes, those were his Phantoms, and it should have been him leading the way.

In the meantime, the Brown family was on tenterhooks. If Eric left the Royal Navy, what would he do? 'He was like a bear with a sore head,' remembered Glenn. Nothing was right and the sailors at Lossiemouth felt it too.

Unsure where his future lay, Eric frantically started looking for new opportunities in aviation outside the Services. With the experience of commanding a naval air station, and managing a population of 6,000, Eric felt well qualified to apply for the job as manager at Teesside International Airport, the former RAF Middleton St George.[4] He flew down for the interview in a Sea Vampire trainer, using the flight as a 'currency check flight' and navigation training exercise. Perhaps it was arriving

4 This airport is now called Durham Tees Valley Airport. It had closed as a fighter station in 1963.

in flying gear, or perhaps it was a clash of personalities with the local council's team, but he did not get the job. This was yet another blow to his pride.

However, the newly created job of chief executive of the British Helicopter Advisory Board soon caught Eric's attention. This industry group had been set up in Redhill, Surrey, with the support of the Civil Aviation Authority and Bristow Helicopters. Eric had first learned to fly a helicopter in 1940 and, as we shall soon see, had always been fascinated with the machine. Moreover, Alan Bristow had even been his messmate during their naval days. They were not bosom pals, but at least one knew the measure of the other, and Bristow knew Eric was not only a celebrated aviator, but also very thorough and professional. Thanks to this, Eric got the job.

On 11 March 1970, fittingly in Scotland, Eric took his final naval flight in a helicopter, the Westland Whirlwind HAR 7, with Petty Officer Wiggins as his crewman. It was snowing, and, as you might expect for this last naval flight, it was not without its share of drama.

Eric's crew also included a newspaper reporter, who was on board to get an airborne view of the heavy snowfall. At 800 feet, a loud bang alerted Eric that his piston engine had failed. He quickly assessed the situation but found the ground features impossible to distinguish in the deep snow. He did, however, notice a wire fence as they were quickly descending. Bringing the helicopter down, he proceeded to hook his tail skid to the wire and made a safe arrested landing into a snow drift. This was the last of Eric Brown's twenty-three major aircraft accidents, and, as he had done many times before, he had cheated death yet again. The reporter was also able to take pictures, which were subsequently carried by newspapers throughout Scotland.

It is interesting to reflect that Eric managed to log air accidents both on the day he left Farnborough and the day he left Lossiemouth. This might just be coincidence, but it might also be that subconsciously he willingly put himself in danger so he could prove his worth and show he would be missed. If all else had failed, it was almost his last chance to prove his greatness, particularly with a reporter watching. This might be overly cynical, but it certainly seems to fit Eric's character, which never missed an opportunity to show off to VIPs.

While Eric's dreams of achieving flag rank in the Royal Navy had been dashed, a second career now beckoned. Eric had much to look back on, reflect on and feel satisfied about. The foundling on an Edinburgh railway platform had become an aviation legend. But now it was time to start the next chapter of his career working with helicopters, and in doing so return to a journey that had started over thirty years before, thanks to an old friend: Hanna Reitsch.

PART 5
Introspection and Reflection

A friend for life. Neil Armstrong and Eric hit it off from the first moment they met at RAE Bedford in 1970. Both were naval test pilots, and both had a strong urge to go faster than anyone else. The relationship held through various lecture tours and gatherings of test pilots in Britain and America.

27

Spacemen and Helicopters

From high up in the gods of the Deutschlandhalle, a teenage Eric watched on with a keen sense of anticipation. It was 18 February 1938, and 25-year-old aviatrix Hanna Reitsch was about to herald a new era of flight: the helicopter.

Over the previous year, Reitsch had been intimately involved in the helicopter development programme, including undertaking endurance and speed trials. On 18 October 1937, the machine was also demonstrated to Colonel Charles Lindbergh, the solo Atlantic flyer and Nazi sympathizer, who was entranced by the concept of vertical flight. Now it was time to show it off for the first time to the German public, as well as the world.

However, as the large crowd looked on at the exotic machine before them, there was a problem. The three giant rotor blades were whirling endlessly without seeming to gain traction. It seemed the helicopter's single Bramo 7-cylinder air-cooled radial reciprocating engine was struggling to find enough air to develop the 160 hp necessary to lift the flying machine vertically off the ground. Something needed to be done, and quickly.

With a clear shout of authority, a Luftwaffe engineer ordered that the great doors of the Halle should be opened. This did the trick. The resultant rush of fresh air gave oxygen to the engine, and the flying machine gradually climbed into the air. Inside the cockpit, Reitsch was in complete control. She manoeuvred the machine, pivoted it 360 degrees, then, with two small bumps, landed it gently on the matted floor 'as gracefully as a bird'.

Eric, along with the crowd, could not believe what they had just witnessed. As Reitsch took in the cheers and rapturous applause, the newsreels were there to capture it all, spreading the images worldwide and adding to the Nazi propaganda drive about the new Germany's aeronautical prowess. Demonstrated by Reitsch on fourteen successive evenings, the show proved beyond a doubt that controllable vertical flight was possible. From this moment on, and despite all his test pilot success, Eric remained enraptured by helicopters.

It was therefore surprising that it was not until March 1945 that Eric took his first helicopter flight, when he and Squadron Leader Tony Martindale travelled to Liverpool to sign for a pair of Sikorsky helicopters.[1] On

1 Called Hoverfly in Royal Air Force parlance and Gadfly by the Royal Navy, but simply called R-4 by the Americans. Veteran John Fay, a contemporary of Eric at the STU, speaks of the R-4 as being the most

arrival, they had been expecting to have at least half a day's instruction, but without an instructor available, they were instead tossed an instruction manual by a burly USAAF crew-chief and told, 'That's all there is, Bud.' Martindale famously said that seeing the pilot's notes for the first time – essentially the operating manual – was like 'reading one's own obituary'.

The next day, after flicking through the manual, the two intrepid aviators managed to fly cross-country, over 200 miles, to the maintenance unit at Andover. It was not until a few days, and a few flights, later that Eric finally attended a short course on helicopter handling with Squadron Leader B. H. Arkell. After his escapades the following day Eric was, of course, a natural. Within hours, Arkell had him hovering, then sent him on his 'first' solo just three days later. The antics of 4 March were apparently rather unofficial in nature.

In his time at Farnborough, Eric carried out further helicopter trials at RAF Beaulieu, including research into rotor downwash and the dreaded 'vortex ring',[2] which is an inherent issue with all single-rotor helicopters. Thanks to this, the boffins at Farnborough started

---

challenging yet enjoyable aircraft which he flew in a six-year career. Fay was one of those on the first naval course, immediately after Eric.

2 A flight condition in which a helicopter that is receiving power from its engine(s) loses main rotor lift and subsequently experiences loss of control. And it tends to crash!

to understand the nature of the aeronautical dynamics involved in rotary-wing flight.

Such was Eric's obsession that Lynn Brown would often speak about her husband sitting in their garage at Farnborough, in total darkness, using two sticks – a broom handle cut into two – and two blocks of wood on the floor, so he might get used to the collective and cyclic responses needed to master helicopter controls. He demonstrated the agility of the helicopter to anyone who wanted to see and managed to convince the Air Ministry to release some Gadflies to the Royal Navy, so that a training school might be set up. In 1947, he even made the first cross-country helicopter flight while totally in cloud, after practising flying blind.

At Whitehall, Eric was also an early advocate of helicopters on aircraft carriers, as he thought that they were perfectly suited to anti-submarine warfare. The machine's ability to hover at a low level meant it was perfect to deploy 'dipping sonar', where a hydrophone is lowered from the helicopter into the sea and can track underwater enemy vessels. The American Sea King helicopter had this system, as well as the range and capacity to carry lightweight torpedoes to become a hunter-killer. Naval cost-cutting by the government at this time sadly meant that this was a non-starter.

At Brawdy Eric also came to see just how useful helicopters could be in remote areas like west Wales.

He predominantly flew to help carry supplies to rocky islands and lighthouses, which were impossible to reach by aeroplane, and was also involved in many helicopter rescues at sea.

Though Eric set many records, he was beaten to the first shipborne helicopter landing on a British warship by Lieutenant Alan Bristow. Bristow became a pioneering legend, first in British helicopter aviation and then internationally. The company which still bears his name provides, for example, the search and rescue services around the British coastlines. Both Eric and Bristow could be irascible characters, but they rubbed along well enough in business.

Bristow may have been first aboard a warship, but Eric was pleased to scoop the first landing of a helicopter on a merchant ship, MV *Dagestan*, and to also land a helicopter on a British aircraft carrier in the later Sikorsky S-51 Dragonfly.

There's no doubt that Eric loved helicopters, but if asked in 1945 if they were to have been his second career, he would have laughed at such a suggestion. He was after all one of the world's foremost test pilots, with the aviation world apparently at his fingertips. But in 1970 things had changed, and for the next two decades, Eric Brown would be closely associated with every aspect of commercial and civilian rotary-wing flying in Britain, Europe, the Commonwealth and even in America. He would build friendships and create a following which few aeronauts have ever managed. Far from the slow march to retirement, these were some of the most exciting years of Eric's life.

On 9 April 1970, Eric started his new job with the British Helicopter Advisory Board. Despite his many ground duties, he still couldn't resist the opportunity to get into the air. So, when he was asked to undertake some trials work at RAE Bedford, Eric literally leaped into the cockpit of a Wessex helicopter. This is how, on 22 June 1970, he came to meet Neil Armstrong, the man universally known as the first man on the moon.

At the time, Eric was engaged in trialling the new Microwave Aircraft Digital Guidance Equipment, abbreviated to MADGE. This kit was destined for the North Sea helicopter operations, to support the Shell-Mex drilling platforms. The oil company needed to have all-weather and day/night capability for its helicopters, with equipment small enough and portable enough to be carried out to the rigs. This would be an area where Britain would show technological and operational leadership. Alan Bristow's vision of offshore helicopter operations in all weathers, servicing the oil rigs of the newly developed North Sea, would be a major driver in these developments.

The MADGE experiments required the whole runway at RAE Bedford to be clear for several approaches and landings. The only other aircraft flying that day was the Handley-Page HP 115 experimental delta-wing aircraft, part of the British work on low-speed delta-wings which originated with the Germans in the Second World War. The original German plans were taken to Farnborough by Eric's old friend Dietrich Küchemann in 1945 and were now being used for Concorde. It was this delta-winged aeroplane that Neil Armstrong had come from America to fly.

'A Land Rover came to dispersal to pick me up,' Eric remembered, 'and when I jumped into the back, there was a young American with a naval crew-cut hairstyle who stretched out his hand and said, "Hello Sir, I'd like to introduce myself, I'm Neil Armstrong,"' and a firm handshake followed. 'My sort of man,' thought Eric.

It was early afternoon, and the mess at Bedford was deserted, allowing the two test pilots to exchange views and personal histories. Armstrong said that he knew of Eric from his days at Patuxent River and the work he had done to make safer naval aviation aboard carriers. They talked about current flying, but Eric did not ask Armstrong about his voyage to the moon. He recalled that 'Armstrong said very vehemently that none of us [who went to the moon] went up there and came back the same person. I'd heard rumours that he didn't like talking about it and he was a bit cagey.'

Eric was actually one of the few who was not that

interested in Armstrong's stories of space. When asked if he envied Armstrong going to the moon, Eric would always reply, 'Why would I want to fly in a machine controlled by computer . . . that's not a pilot's job . . . Neil achieved far greater success closer to earth.' It was this subject in particular that Eric was certainly very keen to discuss.

On 20 April 1962, Armstrong had flown the X-15 rocketplane, then the fastest machine on earth, and in which he had set a world record of flying to 207,000 feet above mean sea level. When asked about his greatest aviation-related regret, Eric admitted, 'If there is one aeroplane I would really like to have had in my log book, it was the X-15. In fact, it is probably the only one on which I missed out.' Eric was perhaps vindicated in his choice of the X-15 as his 'ultimate aeroplane', when Armstrong told him that he'd 'sooner be on the X-15 project than his previous lunar [Apollo 11] one'.

From their first meeting, Eric's relationship with Neil Armstrong became one of the most enduring of his life. 'He was one of my best friends for more than fifty years,' Eric would say when asked about their friendship. 'As Neil said, "we were both naval aviators, rocket ship pilots, test pilots and had family roots in the Scottish Borders – good reiver stock."' The Armstrong family hailed from Langholm, just over the border from Carlisle, and Eric's adoptive family roots were in Langhope, the hamlet above Galashiels. Eric's relationship with Armstrong was another of mutual admiration and kindred spirit. They also shared the same determination and even bloody-mindedness.

Such was their relationship that in 1984, after Eric was elected an Honorary Fellow of the Society of Experimental Test Pilots, an American-based group of the best pilots and engineers, the two great names in aviation began a series of lecture tours, speaking together in America and Britain. In doing so, Eric holds the record for the number of invitations to speak at the Smithsonian Air and Space Museum in Washington, DC – including the Lindbergh Memorial Lecture – seven times, beating Armstrong's five times. In later years, Eric and Armstrong corresponded and spoke on the telephone regularly. In 2010, they even sat down and wrote respective lists of close calls, which became a classic book, with Eric's words and Michael Turner's exquisite art.[3]

In due course, old age and opportunity reduced the relationship to Christmas cards and the occasional telephone call. Both were, however, recognized by the Guild of Air Pilots and Navigations in October 2006 and both received honorary doctorates from the University of Edinburgh in 2008. The following year, they were

3 *Too Close for Comfort: One Man's Close Encounters of the Terminal Kind*, Blacker, 2015.

also both at the Centenary of Flight celebrations in Los Angeles, wowing the crowd with their respective stories of aviation.

The last meeting between the two was very special. In March 2010, Eric was telephoned by the US Embassy and asked if he could make himself available for dinner that evening in Dover Street, in London's West End. Eric had no idea what this was all about. Collected by an Embassy car, Eric was transported to London, where he was met by a member of the Embassy's defence department in uniform. 'What's this all about, then?' Eric asked the smartly dressed Colonel. 'I am not at liberty to tell you, Sir,' replied the Colonel, 'but please come this way.'

After entering the restaurant, and being shown to a private room, Eric's jaw dropped in surprise. There to meet him was his old friend Neil Armstrong, as well as Eugene Cernan, the last man on the moon; Bob Gilliland, the principal test pilot for the world's fastest aeroplane, the SR-71 Blackbird; James Lovell, commander of the ill-fated Apollo 13; and Britain's Wing Commander Andy Green, the world land-speed record holder. They had all been brought together by Armstrong in a belated ninetieth birthday treat for Eric. To Eric, such a gathering of aviation greats was truly manna from heaven.

Meeting the likes of Armstrong was certainly a thrill, but Eric's primary task at the British Helicopter

Advisory Board (BHAB) was to concentrate on devising safe helicopter operations. In this he helped to develop the framework under which helicopters operated, both offshore – North Sea oil exploration was beginning to be a major user of helicopters – and onshore, where high-voltage electrical pylon lines and underground pipelines were being inspected from low-flying helicopters (in places that traditional fixed-wing aircraft could not safely operate). Use of British helicopters overseas for crop spraying and pest control also fell under Eric's list of responsibilities, as the BHAB was seen as the 'gold standard' for new helicopter operations at a time when Britain still led the world in aeronautical expertise.

One of Eric's initiatives was to support public transport helicopters, operating from city centres to major airports, or between destinations which were not served more conventionally by road, rail or airports. He believed that the success of the US Marine Corps, and the US Army in transporting troops and supplies around the battlefields of South-east Asia, was pointing the way to civilian transport uses. If helicopters could be used safely in war zones, then why not, he would argue, in city centres? For public transport hubs, Eric and the BHAB team advocated for Battersea Heliport and other proposed sites in the City of London, including a rooftop heliport, and a flying pad near London Bridge. It helped that the then-state-run British European Airways

Helicopters Ltd was keen to develop helicopters for onshore flying and, particularly, for offshore work.

Helicopter medical evacuation, which started in Korea, and was fully developed for the military in Vietnam, also spurred Eric to take up changing the policy on air ambulance, police and even airborne firefighting tasks for helicopters. Public services, he believed, should be encouraged to use helicopters for the emergency services, irrespective of the costs. In Europe, he travelled to Switzerland and Austria to see aerial logging, pylon erection on mountainsides and then, in Cornwall, to watch Trinity House light-keepers being transferred by helicopter to the last of the manned lighthouses. This, of course, helped pave the way for many of these things in Britain, which today we take for granted.

Although he had always dreamed of becoming an Admiral in the navy, or to continue to be a test pilot, life with BHAB at Redhill was in many ways just what Eric needed at this stage of his life. He had many friends in Surrey and Sussex, where he chose to live in a private development at Copthorne, on the border of the two counties. 'Carousel' was his first home bought with his own money – his annuity from his service with the Royal Navy – and it proved ideal. The house was modern and convenient, had some entertaining neighbours and for Lynn represented a permanence which she undoubtedly craved. Two main features of 'Carousel' always stood out to visitors – and there were many of those. First was the oak beam, which had been built in by the first owner, a shipbuilder, and the second was the bar underneath it, where many a gathering was held. Bizarrely, there was also a combined radio/toilet-paper holder.

First and foremost Eric was preoccupied with safety at the BHAB, but there was still plenty of opportunity to have some fun. The BHAB was a great supporter of the Helicopter Club of Great Britain (HCGB), which, among its many offerings, fielded international teams for competitions in the Soviet Union, Poland, Germany and France, as well as its annual competition in Great Britain.

Eric, with his excellent knowledge of the German language, was much in evidence at the First World Helicopter Championship, held at Bückeburg in September 1971. Germany had already held national championships in 1963, 1965 and 1969, so the basic format of such competitions had been established. To everyone's surprise, not to mention Eric's, Hanna Reitsch appeared as Captain of a Hughes 300, with British co-pilot Mandy Finlay.

The second international event under Helicopter Club of Great Britain auspices was at the Army Air Corps Centre at Middle Wallop, Hampshire in July 1973. This time things became far more serious as the Americans put everything into winning, with Bell Helicopters providing three days' training, all expenses paid, and also chartering a Bristow-operated Bell 47G2 piston-engined helicopter.

Eric seems to have approved of the dedication of the US team, and the way in which the Whirly-Girls, an America-based group of helicopter-flying women, coped with their helicopters and the British weather. It was not just Eric they impressed – they were even sent a telegram of good wishes by Hanna Reitsch. Another notable guest in attendance was Prince Philip, Duke of Edinburgh, who, invited by Eric, flew himself to the event in a Queen's Flight Wessex helicopter from RAF Northolt.

The result of British hospitality, and army organization, was a letter to Eric, dated 7 August 1973, from Ralph Alex of Sikorsky Aircraft, and the president of the FAI (Fédération Aéronautique Internationale) rotorcraft section: 'Your tireless, cheerful and indefatigable efforts were surely a major ingredient to this Hover lovers' six-day "rotor fest".' Eric was delighted, and so was Lynn, who was mentioned too. On this occasion, the Soviet Union won, the Royal Air Force's Central Flying School came second, and the Austrian Air Force took third place.

The Third International Helicopters Championship was held in the Soviet Union, not far from Moscow, and attracted the same teams as before. This time around the competition was far more fierce, as Cold War politics played a big part. It was clear that the Soviet Union and the United States saw the competitions as an extension of the Cold War and put considerable military and political effort into winning. America, too, had commercial interests in ensuring that its constructors, such as Bell and Sikorsky, won. Eric wasn't convinced about state-supported teams, as he wanted the 'little man or woman', as he put it, to triumph.

There were, however, some benefits to this. Eric particularly remembered arriving in Moscow as one of the judges. At the airport, he was whisked through the diplomatic channel and taken straight to a large Zlin limousine. When he protested that he was only one of the judges, and not the chief judge, who was American, the Soviet official simply stated: 'You are English [sic]; you were allies against the Nazis when other countries stood by and just watched.' Eric never forgot those sentiments.

Yet, as always when it came to aviation matters, Eric was determined to have his say. After the event, Eric corresponded with Otto Rietdorf, the German competition director from the German Aero Club. Eric was in favour of modifying the World Helicopter Championships, starting with the programme, reducing the number of teams to one team of five crews, to reduce the 'professionalization', as he saw it, of the Soviet and American teams. He was also ahead of his time in being against having distinct male and female teams in the events and wanted mixed crews. He felt that spectator appeal was important too, so he wanted the timed arrival, and navigation exercises, to be withdrawn, as they favoured the host nation, who would know the terrain. Most importantly of all, he was also growing concerned at how the Cold War was overshadowing the championship. The Americans were overly concerned that the Russians – the Soviets as they still were – were winning competitions in basic, piston-engined designs and somehow beating the latest US technology. In the end, the US Army's dedicated team at Fort Rucker[4] were training for weeks as if their very careers were at stake, and this took the fun out of the Championships, while the Soviets also used gamesmanship to get any advantage they could.

While Rietdorf was happy to listen, Eric was disappointed that not all of his suggestions were taken up. 'The trouble was,' remembers Elfan ap Rees,[5] 'that Eric always wanted to be in charge, whether it was the HCGB board of directors, the competition or things to do with helicopters in general. Just because he had flown the

4 Fort Rucker is the alma mater of US Army aviation. On a good day, it is said there are more helicopters flying in the circuit there than equip the whole of the British Army.

5 Editor of *Helicopter International*, HCGB stalwart and founding director of the International Helicopter Museum at the former Westland plant at Weston-super-Mare.

early Sikorsky Gadfly and other helicopters in the Fleet Air Arm, he thought he knew all about helicopters.'

Sadly, as had happened so often in Eric's career, he found it impossible to hold his tongue. The HCGB, like so many clubs, only works where there is a common purpose. When there are big personalities, and big personality clashes, trouble is never far away. So, it was to be with the HCGB at Castle Ashby, Northamptonshire in 1986, where after years of tension, things finally came to a head.

During a prize-giving dinner, Robert Pooley, a leading member of the HCGB, remembers that Eric was upset that a junior officer was seated next to Sergei Sikorsky, son of the great helicopter pioneer Igor. Although the young officer had won the trophy for the best individual performance, Eric clearly felt that he was still undeserving of the honour. Once more, Eric's ego got the better of him. Just as in his time training pilots in the Royal Navy, he refused to accept that anyone, other than a fellow test pilot, was on his level. He had to be seen to be the best and wanted everyone to know it. By now he was sixty-six and had enjoyed a glittering career, and yet still he had a chip on his shoulder.

Unsurprisingly, Eric's attitude led to much rancour between him and others on the committee. He accused them of favouritism, while they were also unafraid to hit back at what they believed was uncharitable behaviour.

'Eric never liked being challenged,' said ap Rees, adding, 'It all ended in tears.' As a consequence of his behaviour, Eric was eventually forced out of the HCGB.

Soon after, his time with BHAB also came to an end when he retired in 1987. There was no such upset in this case. It was merely time to be with his family and enjoy his retirement while he still had his health. Due to his achievements and service, the Board honoured him by instituting the annual Eric Brown Award for the year's most outstanding contribution to helicopter aviation. Eric had made a considerable, positive impact on the helicopter industry, increasing safety and lobbying government for recognition of an industry then in its infancy. His pioneering work on elevated helipads, and their implementation, set an international standard which is still used today. Robert Pooley also threw a considerable 'olive branch' into the mix by awarding Eric a special miniature naval ceremonial sword to recognize his 'outstanding achievements'. They parted friends, yet Eric remained convinced there was a vendetta against him at the HCGB until his dying day.

Eric had enjoyed a wonderful career and had achieved everything and more he could ever have dreamed of. It was now time to enjoy his retirement, and all the accolades that came his way. And yet Eric couldn't rest. In his eyes, it was now time to set the record straight.

'You make James Bond look a bit of a slacker' was the verdict of Kirsty Young, who had chosen Eric's life story for the 3,000th edition of the BBC Radio programme *Desert Island Discs*. Even in his nineties, Eric was a charmer and had Ms Young eating out of his hand. He told of his youth, created a few myths and generally had a most excellent time. It remains the classic interview.

28

National Treasure

For the first few years of his retirement, Eric and Lynn did just what other couples in their position would do. They travelled the world, saw old friends and made new ones, such as engineer Eddie Albrecht and his wife Sue. Eric also attended plenty of professional gatherings, including giving the address at a memorial service to Sir Frank Whittle at Westminster Abbey. While aviation matters still dominated their social calendar, perhaps the highlight was Eric taking Lynn to Sarasota, Florida, in order to celebrate her seventieth birthday. On the white sands, they toasted their past and looked forward to their future.

Despite his retirement, and few opportunities to fly, Eric was also able to undertake his last flight, which is marked in his logbook. Typically, it did not go smoothly. In 1996, he had been awarded a lifetime honorary membership of the Society of Experimental Test Pilots in Los Angeles. The team looking after him suggested that rather than go 'civ-air' back to Washington, DC, where he had a speaking engagement, he might like to take advantage of a spare seat in a military Beechcraft King

Air, known to the US Navy as the UC-12B. Needless to say, Eric jumped at the chance.

However, while at Kansas City, on a refuelling stop, the pilot-in-command, one of the US Navy's most senior test pilots, received a compassionate call that his father was dangerously ill. The pilot elected to go home via a faster civilian airliner, and, to Eric's delight, he was asked if he wanted a final trip across America as pilot-in-command. 'I know your background,' said the American naval aviator, 'and the guy in the right-hand seat is a naval commander who knows the aeroplane backwards, so off you go.' Eric eagerly grasped the opportunity. 'I landed successfully,' he remembered, 'and as I taxied in, my wife [Lynn] came out to meet the aircraft and saw me up the front and I can still see her face.' Lynn's greeting was simple: 'What the hell were you doing up there?'

These were truly 'golden years', and life was finally there to be enjoyed. However, in the late 1990s, Eric's whole world was turned upside down. After a few months feeling unwell, Lynn was diagnosed with terminal cancer. She died on 28 September 1998, and for Eric it was traumatic. They had been together since 1941, and had been married for almost all that time. She was his anchor and was a fighter in her own right; a lady of immense strength of character and determination. That final morning, the couple were found by the daily help, Eric still cradling her body several hours after she had passed away.

Following Lynn's death, there was a period where Eric seemed lost without her. It was almost as if he was an orphan again, trying to find his place in the world. While his beloved wife had passed away, many close friends and contemporaries started to join her, including Lewis Boddington, the towering presence of British naval aviation engineering, and Gwen Alston, a private pilot and test observer at RAE with whom Eric had a special working relationship during the Second World War, when they jointly examined why so many Fairey Barracuda torpedo-bombers were crashing into the sea after weapons release.

At this time the test role at Farnborough was also being reduced. It had first merged with Bedford into the Royal Aerospace Establishment (1988), then the Defence Research Agency (1991) and finally the Defence Evaluation and Research Agency (1995), so defence research and development could then be privatized. With America and France still investing in new aeronautical technology, Britain seemed to be going the other way. This upset Eric. He had risked his life countless times, and now it seemed there was a real danger that all he had achieved might be forgotten, just as he approached his final years. In response to all of this, he became determined to share his story and expertise and grasped any opportunity that came his way.

Firstly, he made himself readily available to the

Farnborough Air Sciences Trust, to ensure that the science and technology developed by the skilled men and women of the Royal Aircraft Establishment would not be forgotten. This also ensured that his own legacy would remain intact. Every year he also presented the annual 'Eric Brown Award' from the British Helicopter Advisory Board, of which he had been chief executive. This eventually led to an invitation to lecture in the United States, including at the Charles Lindberg Memorial Lecture at Langley in the IMAX theatre.

When offers arrived for Eric to work as a consultant, he eagerly accepted. In 1998, a Defence White Paper from the government called the Strategic Defence Review spoke for the first time about bringing large aircraft carriers, with significant fixed-wing strike capability, back to the Fleet. Two teams were created by the industry – at BAE Systems and Thales – and both engaged Eric for his experience on the CVA-01 project and about aircraft operating issues. Eric later flew to America to consult with Lockheed-Martin, the design authority and manufacturer of the F-35 Joint Strike Fighter – although he could not understand why the British government did not 'bite the bullet' and put a proper catapult on the new carrier, HMS *Queen Elizabeth*, so it could inter-operate with the US Navy. In his later years, he had lost none of his scathing thoughts about politicians, while he was also unable to hold his tongue at what he saw as amateurs interfering with things they knew little about.

In Toulouse, Airbus also recognized Eric's undoubted talents and experience in specific flight configurations. When the giant Airbus A380 was undergoing trials after its first flight in 2005, the test and evaluation team asked Eric to join one of the test flights, which was exploring the transonic handling of the airliner in a shallow dive. Although Eric was now in his mid-eighties, there was no one better qualified in Europe to sit in the right-hand, co-pilot seat and advise the team as the aircraft reached the critical Mach numbers. For Airbus, with its Franco-German heritage, Eric's long association with jets, the sound barrier and test flying meant he was held in high esteem by the international test pilots. Many of these aviators had trained at the Empire Test Pilots' School at Boscombe Down, where Eric's original test flying methodology was still taught.

Television companies, as well as aviation historians, also realized that Eric was not only knowledgeable, but also very approachable. On 7 August 2005, the BBC interviewed Eric for the documentary *Hunt for Hitler's Scientists* – after all, even in 2005, there were few alive who had met, interrogated and befriended them. Eric's contribution was to talk about the Messerschmitt Me 163 rocket fighter, which he was so proud to have found, flown, mastered and survived; the only non-German pilot to have done so. The Me 163 was so important to the postwar breaching of the sound barrier, and subsequent high-speed flight developments, that Eric was always in demand to talk about it.

Such was his standing that in 2014, Eric was invited on to the 3,000th edition of the BBC Radio programme *Desert Island Discs*, where he was described by fellow Scot presenter Kirsty Young as a 'remarkable dare-devil' who 'makes James Bond seem like a bit of a slacker' and is 'a real-life hero'. As an avid big band fan, Eric chose as one of his songs 'At Last' by the Glenn Miller Orchestra. This choice was unsurprising, as Eric was not only a Glenn Miller fanatic, but had a unique connection to the band.

Before Glenn Miller's fateful trip to Britain in late 1944, Eric and Lynn had secured tickets for his last public performance on 14 December at Milton Ernest Hall, on the outskirts of Bedford. It might seem like a strange venue for the world-famous Glenn Miller Band, but Bedford was a 'liberty town' for US servicemen from the surrounding US Army Air Force and other camps. The BBC had also been evacuated to the town in 1940, and Lynn had contacts there. Thanks to Lynn, who was a celebrity in her own right as a BBC radio singer, the two Browns met Glenn Miller before the show. Previous visitors in the days before had included David Niven (about to go on active service in the Ardennes), Bing Crosby and Bob

Hope. With Miller's customary good humour and exquisite manners, he invited Lynn to sing with the band and, seeing Eric's reaction, invited him to play the drums.

After a memorable evening Miller told the happy couple, 'You can both come back any time.' Eric and Lynn would have done so, but fate intervened. The following day, Miller was due to fly to Paris from nearby RAF Twinwood Farm to 'make arrangements' for the band to play on Christmas Eve. En route, the aircraft disappeared over the English Channel in atrocious weather, with all on board lost. The reasons for the crash remain a mystery, but Eric always maintained that the flight should never have taken place and he was surprised at Miller's impatience.

In Eric's later years he also gave talks on 'Britain's Defence in the Near Future' and warned about defence cuts at a time when the United Kingdom needed to stand up to Russia. His speech was widely reported, and at the time of writing seems prophetic after the Russian advance into Ukraine. He was also unafraid to share his views on Scottish independence and Brexit, being quoted at length in national newspapers, campaigning for the union, while also being an ardent Leave supporter. He had never forgiven the German and French negative attitudes towards Britain when it had wanted to join the EEC following the war and believed Britain should stand on its own two feet.

Throughout all of this, Eric was showered with awards. The Guild of Air Pilots and Air Navigators gave him the Award of Honour for outstanding lifetime contribution to aviation in 2005; the award was given jointly to Eric and his old friend, astronaut Neil Armstrong. In 2014 came the unveiling of Jenna Gearing's bust of Eric at the Fleet Air Arm Museum at Yeovilton, which had been commissioned by the Museum's Society of Friends. The bust depicted him dressed for flying and was positioned with the Museum's Martlet fighter marked in his colours. It was an accolade that truly touched Eric.

Eric also greatly appreciated and respected the University of Edinburgh. The University had allowed him to leave after two years of study to join the Royal Navy and left his degree frozen for 'the duration' – as the Second World War was often referred to. He finally gained his suspended degree in 1947, but in June 2008, he travelled to Edinburgh, accompanied by son Glenn and new daughter-in-law Ros, to be awarded an honorary doctorate. He appreciated this award more than almost any other, and it came at a time when his morale needed to be boosted because he was in pain. He had been diagnosed with an enlarged prostate, which, although not acute, needed surgery. It certainly meant a pause in his lectures as he recovered at home.

After spending some time recovering, Eric returned to the 'circuit', and a trip to the past, with a visit to

Lossiemouth as the guest of honour for the station's annual officers' mess Battle of Britain dinner. The Station Commander, Group Captain Al Monkman, pulled out all the stops to take Eric back in time, first to 1940, when he had begun his naval career, and then to the period between September 1967 and March 1970, when he was the Captain of the then Royal Naval air station. Monkman recalled that Eric 'provided many stories and photographs to help build up the station museum over the last few years'.

Unsurprisingly, as Eric's profile grew, he was inundated with letters and telephone calls. He prided himself on answering everyone, particularly if he could set the record straight on some aviation matter. If some correspondents were just time-wasters, they would get a short, if courteous, reply, but with others, with real shared interests he would develop close relationships, even if just by post. Many of the correspondents were German, and those exchanges would be conducted in German. Eric even called several contacts in Bavaria and Berlin on the telephone on a regular basis to 'keep my language skills up to date'.

He also became pen-pals with an eleven-year-old schoolboy from Dorset called Eachen Hardie. Like many young boys, Eachen wanted to be a pilot, but not just any old airline pilot – he wanted to be a test pilot and had identified Eric as his hero. He collected pictures and cuttings about Eric and in 2014 plucked up the courage to write to him and ask for advice prior to starting secondary school in Dorchester. Eric's advice was straightforward – become proficient in maths and physics and stay physically fit. Eachen continued to write to Eric and confirmed that he had moved into the top stream for maths and was in the school swimming team. They never met, but Eric was touched that he might still help inspire the next generation of test pilots.

However, while Eric kept himself busy, he also found comfort in a companion again. Jean Kelly was the widow of Eric's friend Lawrence Kelly, better known as 'Red', because of his hair colour. Red Kelly had been a student pilot with Eric at Netheravon in 1940, before commanding Seafire units at sea and being awarded the Distinguished Flying Cross for his exceptional courage during Operation Dragoon in 1944, when the Allies liberated southern France. While Red retired from the Royal Naval Volunteer Reserve to practise law in Birmingham, the Kellys and Browns always made a date of the annual Fleet Air Arm dinner, attending as a foursome. When looking for a retirement home, Jean and Red opted to convert and renovate an old stable block on the Cornish coast at Rock. Sadly, Red died within days of the couple moving to Cornwall, leaving Jean alone, with her closest family at Bristol.

Not long after Red's death, Eric called her one

evening, and reminisced about the old days. He also asked how she was coping as a widow in Cornwall. Jean apparently replied that she was fine but found being a widow rather cramped her previously busy social life. In fact, she must have been quite lonely. Eric was lonely as well. He suggested she should get on the train and come and visit him in Sussex. So, she did.

For the next decade and a half, they made a team for social events, giving each other moral support, and established a close bond. Both loved to travel overseas, so Jean joined Eric for a trip to Bermuda, where he had a speaking engagement about German jets. Within a few years, Jean had heard Eric's stories so many times that she could prompt and remind him, as Eric would not talk from notes.

Eric's willingness to share his past in his later years was only natural. Indeed, anyone in Eric's position would have done the same. After all, he was a renowned pilot, with plenty of thrilling stories to tell. Such a life meant he was always going to be in demand, and why shouldn't he share his experiences, and take the accolades in the process? However, it might also be said that, as time went on, Eric very much had his own agenda.

Even after all these years there was still an intense anger under the surface. While repeatedly heralded as one of the finest pilots of all time, he still felt he had been unfairly locked out of the inner circle all of his life. In his view, lesser pilots than him had climbed the ranks right to the top of the Admiralty. He still judged someone's worth on their ability as a pilot and certainly didn't reflect that he had often brought much of his misfortune on himself. At some point, it seems he realized that if he was going to talk about the past, he also had the opportunity to set the record straight and to craft his own legacy, one which didn't always marry with the truth.

It certainly helped that not many contemporaries were able to question or correct him. Most were dead, while other events relied solely on Eric's memory. Some stories might have been the result of harmless embellishment, or the effects of old age, but others, as we have seen, can't so easily be explained. Some involved dates that didn't match, aircraft he claimed to have flown, or arguments with rivals he won, or should have. Others involved flights of fantasy, particularly involving his family background, and his father's fictitious RAF career, which he only chose to share for the first time in his 2006 reissue of his book. Such things can't easily be brushed away, other than an attempt to write his history as he wanted it to be. It is clear that Eric needed to keep the hidden family history concealed for the whole of his life. He shared nothing with his family or with his closest friends.

Despite this, Eric was grateful for opportunities

which allowed him to revisit the past and reflect. Perhaps the one which pleased him more than any other came in May 2012, when he was invited to return to Germany, in order to address Luftwaffe Aircrew Veterans at Berlin-Gatow airfield. In so many respects, Germany was not only a country which had helped inspire him to become a pilot but it was also where he had spent many of his happiest years. He never forgot the atrocities of the war, particularly Belsen, yet he truly loved the German people and always felt at home there. The opportunity to return one last time was too much to resist.

As he drove through Berlin, Eric saw that much had changed. Berlin was no longer divided, following the Berlin Wall being brought down in 1989 and the collapse of the Soviet Union in 1991. With the war a distant memory, Germany had rebuilt its cities and had also returned to being an economic superpower, once more at the forefront of technology. It pleased Eric to see the hateful Nazi rhetoric consigned to history and Germany now back where it belonged.

On arrival at Berlin-Gatow airfield, Eric, dapper as always, in his blue Spitfire Society tie, navy-cut blazer, grey flannels and highly polished shoes, mounted the stage and looked out to the crowd. With the sun setting behind him, he felt at home. He was among aviators in a country with which he had been closely involved for most of his long life. Although his audience spoke good

English, Eric chose to start in German. Rather than talk of his time as a test pilot during the war, for this occasion he chose to concentrate on telling the tales of flying fabulous flying machines in the Germany of defeat and ruin in 1945. There were treasures to be found among the carnage of six years of war, including new technology that stunned the Western Allies. German aeronautical technology was a decade ahead of what was thought to be leading-edge in Britain and the United States.

That night, Eric reminded his audience of Germany's daring developments in the use of rockets, jet engines, wing and flying surface design, and the ability of a brilliant, hard-pressed group of engineers, designers and technicians to improvise, despite what seemed like the best efforts of Adolf Hitler to interfere with the process and direction of research and production. Notwithstanding Allied bombing of key factories and installations, the Germans had mastered much of the technology which still eluded the best brains in industry, and at establishments such as Wright-Patterson in America and Farnborough in England. Eric firmly believed that the technology brought back from Germany in 1945 gave Britain fifty years of aerospace success – he would always add, 'despite the best efforts of politicians'. For this he earned a standing ovation. The boy who had once visited Germany as an aviation enthusiast was now being applauded by some of its greatest aviators. It was an emotional moment for Eric, and one he was glad to have experienced, as not long after he would have found such a trip impossible.

In 2015, following a reception at the House of Lords, Eric slipped on the stairs at Westminster underground station and fell more heavily than those present realized. He had in fact cracked three ribs and damaged his wrist. At ninety-six years old, any fall is dangerous, but Eric bravely shrugged off the effects. It was, however, clear to his family and closest friends that, while he tried to hide it, he was in pain. Yet there was one more event he felt he could not miss.

When Eric's friend Alfie Southwell organized his ninety-seventh birthday party at Buck's Club in London's West End, he jumped at the chance to see old and new friends. In typical style, a limousine whisked Eric and Jean from Surrey to the Ritz on Piccadilly, where the manager immediately took steps to upgrade the couple to a suite.

Lunch was served in the dining room overlooking Green Park and Buckingham Palace. Looking out, Eric reflected on the half-dozen visits he had made to royal investitures. There were in fact so many during the Second World War that on the award of his military OBE – which had followed the Distinguished Service Cross, the military MBE and the Air Force Cross – King George VI was heard to say, 'What, you again?'

Guests at the lunch included Eric's favourite photographer, John Goodman, and Flight Lieutenant Tim Dunlop, the Battle of Britain Memorial Flight's bomber leader and chief Lancaster pilot. Dunlop was to be lectured, gently, on the flying and displaying of the Lancaster, as well as being briefed on the technique used to roll one. The champagne flowed and, of course, it was Pol Roger. Eric, like his hero Churchill, had a sophisticated taste for the finer things.

That evening, 120 guests gathered at Buck's, where his pains were forgotten, although Jean ensured that he used a stick to navigate the winding stairs. Eric was joined by esteemed British and American pilots, supporters of the Fly Navy Heritage Trust, as well as the First Sea Lord, Admiral Sir George Zambellas. His first command, as he entered the first-floor bar – where the champagne cocktail Buck's Fizz had been invented – was: 'Winkle, report!'

Eric enjoyed the meal, especially as the menu was selected to assist his challenged dentistry. After a personal letter from David Cameron, the Prime Minister, had been read out, Eric rose and spoke. For thirty minutes, he held the room and never once did he look down at any note. For many it was a reminder of a bygone age, when Britain was at the forefront of aviation technology, and its test pilots were among the bravest in the world. At the end of the dinner, the guests were presented with copies of various books Eric had written – all of which he had signed that afternoon rather than take a nap.

When the end came, it was swift but not unexpected. Within a few weeks of his fall at Westminster, Eric fell again and had to go to hospital. After a couple of days, it was clear that Eric would not be able to fend for himself at home. A little later, on 21 February 2016, he slipped away in the early afternoon. A quiet and uneventful end to a life that had thrived on adrenaline and drama.

On his passing, the Fly Navy Heritage Trust organized a memorial event at the Royal Naval Air Station Yeovilton in Eric's honour, compered by Emily Horton, the last journalist to interview him. In doing so, the Trust rolled out its historic aeroplanes from the hangar, and the Fleet Air Arm Museum also brought out several types across the runway. At lunch, Prince Andrew, the then Honorary Rear Admiral of the Fleet Air Arm, spoke, and after his departure, the invited guests watched a flying display and drank Wadworth's Swordfish beer, topped with navy rum. Eric would have approved.

While there were plenty who were willing to pay tribute to Eric in death, there were also many who would not attend his memorial service. Even after he had passed, Eric still roused deep passions in veterans of the Fleet Air Arm. He had upset many with his attitude over the years, and some were unable to forgive him. Still, his achievements cannot be denied.

In an enthralling career, Eric logged more than 6,000 flying hours in 487 different aeroplanes and helicopters, a certified *Guinness Book of Records* number of which he was so proud. His tally of aeroplanes and helicopters is unlikely ever to be beaten. It is, however, his immense bravery in his days as a test pilot, during and after the war, that will always be remembered more than any other. Not only did he help to develop aeroplanes and aircraft carriers that proved crucial to winning the war, but his escapades in Germany with the Farren Mission proved vital in revolutionizing the British and American aviation industries for the next fifty years. He could be prickly and perhaps arrogant at times, but we are unlikely to see his like again.

In his later years, he might have been guilty of embellishment or fantasy, but does it matter? Is his reputation diminished because his first flight was most likely in an Avro 504K rather than a single-seat fighter, propped on his father's knee? Did it matter that he chose not to share his lineage as an adopted child? That might have helped us to understand his character a little better when he acted out of turn, but nevertheless, he is still a national treasure and an aviator with a unique, unparalleled and unrepeatable story. There will never be another like 'Winkle'. In so many ways, he was our greatest pilot.

Appendix 1: Winkle's Top Ten Aircraft

Eric Brown was able to assess and evaluate over 480 aircraft types in a thirty-year flying career. His unique position as Chief Naval Test Pilot at RAE Farnborough with detachments to Patuxent River Naval Air Test Center and on squadron duties with the Fleet Air Arm allowed him to create a list of his top aircraft.

1. **De Havilland Sea Hornet**. This twin-engined naval fighter was too late to see active service in the Second World War and too early to see service in Korea. It was, however, in Eric's mind the finest machine he ever flew. For him, it had all the advantages of the Mosquito and very few, if any, of the faults – counter-rotating airscrews made for a straight line-up on the flight deck without the inclination to wander towards the island superstructure. Eric's first deck landing was on 16 August 1945, aboard HMS *Ocean* in PX212.

2. **Messerschmitt Me 262A.** The first operational jet fighter had a special place in Eric's heart. He was probably the first Allied pilot to fly it and he certainly relished the sweep-wing design, even if he didn't quite trust the power plant. Eric's logbooks are unclear on the exact number of flights, but there are significant dates, such as 24 June 1945, when at Grove on a cockpit familiarization and engine run the port engine exploded. That did not deter Eric, and he would speak fondly of the 262 and its potential to have changed the war had it been introduced into service sooner.

3. **Supermarine Spitfire Mk XIV**. Eric flew most but not all of the Spitfire and Seafire marks, as well as undertaking comparison flights against Fw 190, Bf 109 and US fighters. His verdict, repeated many times, was that the Mk XIV Griffon-engined Spitfire was the best of the mark and the best piston-engined fighter of the Second World War. He based this appreciation on his calculation of the tactical Mach numbers, which put Spitfire before Fw 190 Dora and ahead of the P-51D.

4. **Focke-Wulf Fw 190D-9.** The Dora was Eric's second most liked and admired fighter of the Second World War. He rated it just below the Spitfire Mk XIV primarily in respect of climb rate and reliability. Eric's first Fw 190D-7 flight was probably on 5 May 1945 in 211016 from Grove to Schleswig. It could nearly have been his last, as he was engaged by Canadian troops as he approached an airfield which had yet to be taken. He celebrated his close shave with a flight in a Me 262B-1a two-seater later that day, probably a check-out with Hauptmann Miersch.

5. **Grumman Bearcat**. This was an aeroplane of sheer exhilaration for Eric. He describes it as a 'crackerjack, the best American piston-engined fighter I ever flew'. It was his favourite machine in the Patuxent River flight test inventory in 1951. He says it compares favourably with the Hawker Sea Fury then in Fleet Air Arm squadron service.

6. **North America P-51D Mustang**. This amazing American fighter with its Merlin engine and exceptional range delighted Eric. He rated it as his third most favourite piston-engined fighter of the war but said that he disliked the laminar-flow wings, which he reckoned reduced the tactical Mach number so important for a fighter. Eric did not have the opportunity to fly the Mustang in combat which might have changed his view.

7. **Supermarine Spitfire Mk IX**. In the spring of 1943, Eric undertook some rather 'irregular' operational fighter sweeps over occupied France with the Royal Canadian Air Force. He flew Mk IX BS428 for nearly two hours on 4 April 1943 and recorded how much he enjoyed the agility of the Mk IX. Two days later, he flew MK Vb EP278 on a rodeo sweep across France and said the difference was clear: as a fighter which he had flown in combat, it was the Mk IX which he preferred. He liked the speed of the Spitfire Mk XI, which was used as a trials aeroplane at RAE Farnborough.

8. **Grumman Martlet**. The robust little American fighter which Eric first flew in 1941 and in which he carried out the first of 2,407 deck landings; a world record. He flew all the four marks available to the Fleet Air Arm and loved the fighter's robustness and ability to withstand damage.

9. **North America F-86A Sabre**. Eric had to wait until he was posted to America to go supersonic. He achieved Mach 1 in a dash in a Sabre at Pax River and thereby broke the sound barrier for the first time on 30 September 1952 in 491308 and took it to Mach 1.12 with Vought XF7U-1 Cutlass 124419 as chase-plane. On 23 November 1951, he had taken the same F-86A-5 to Mach 0.96, the fastest he had flown to date. By January 1952, Eric had amassed 3,739.05 hours of flying as pilot in command.

10. **McDonnell-Douglas F-4B Phantom**. When Eric calls an aeroplane a 'hot ship', one knows it is good. He only flew the Phantom once and then it was American, but he did get to feel the power of this superb carrier aeroplane. He felt justified in having brought the concept into being for HMS *Ark Royal*'s carrier air group even if he did not fly the F-4K variant.

Deserving of special mention:

Bell Airacobra Mk I. Eric used AH574 as his personal run-about and trials aircraft. He pioneered some air-flow trials which would add to RAE Farnborough's knowledge of how a jet might perform on 13 April 1945, used the same machine for flexible deck trials and flew it around Europe until it was condemned as unsafe by Bell Aircraft on 28 March 1946. Eric said the Airacobra suited him 'very nicely' in terms of handling, range and comfort. He would not swap it in for a new model.

Eric also had a soft spot for the **Supermarine Seafire Mk XV**, with which he carried out the first tail-down catapult launch aboard HMS *Indefatigable* on 12 May 1944. In all, Eric logged 2,721 assisted or catapult launchers; a world record.

Two that got away:

The two aircraft Eric would have liked to have flown were the experimental **North American X-15 space-plane** and the ultra-high-altitude and very fast **Lockheed SR-71** strategic reconnaissance machine. One was not possible because he was British, and the other was a little after his time.

Appendix 2: Winkle's Least Favourite Aircraft

1. **General Aircraft GAL/56**. It's clear that this aeroplane was the all-time party pooper of Eric's logbook, and he spoke frequently about it in his later years. Eric once said: 'They don't come much worse than this [aeroplane].' He also believed that the RAE Farnborough flight test programme was a battle of wits, as it was test flying by trial and error – it is the perfect example of 'If it doesn't look right, it won't fly right.'

2. **De Havilland DH 108 Swallow**. Eric describes the Swallow as 'a serial killer' which cost the life of Geoffrey de Havilland on 27 September 1946. Eric flew another prototype with a redesigned cockpit hood which probably saved his life – along with his stature – when he undertook trials in 1948. It would later claim the life of Squadron Leader Stewart Muller-Rowland in 1950.

3. **Fairey Spearfish.** This 10 ton monster was a heavyweight strike aircraft destined for the Pacific, but war ended before production began. Eric was relieved that this 'real old cow' did not enter naval service.

4. **Supermarine Attacker.** From the makers of the world's most iconic fighting aeroplane, the Spitfire, came a naval fighter which just did not cut the mustard. Undistinguished in so many ways and not loved by its pilots on account of the far-forward cockpit and the risks involved with a steel barrier landing on the axial deck of aircraft carriers. Eric first flew it on 2 September 1947 at RAE Farnborough.

5. **Bell Airacomet**. This was one of the contenders for the first jet landing on an aircraft carrier, along with the Gloster E28/39 and the Meteor. Eric rejected all three in favour of the Spydercrab (later renamed the Vampire). In fact, he summed up the Airacomet as dull and ponderous, even if it was the first US jet fighter.

6. **Blackburn Firebrand TF IV**. This large Bristol Centaurus-powered machine was another 'less than perfect' British design which Eric described as 'a disaster as a deck-landing aircraft'. Eric's first encounter with the Firebrand was on 1 February 1943, when he acted as a Deck Landing Control Officer for Commander Denis Cambell, the designated test pilot for the type.

7. **Dornier Do 335A.** This revolutionary twin-engined fighter was perhaps too revolutionary. It was one of the few German aircraft about which Eric would privately offer reservations. This is borne out by the flight of one example from Reims to Merville on 13 December 1945, which resulted in a heavy landing for 240112/AM225 after a complete hydraulics failure. Eric was not flying it but had authorized the German test pilot, probably Hauptmann Miersch, to carry out the ferry flight.

8. **Miles M239B Libellula**. Just one look says it all: not really a serious act of aerial warfare. The Libellula was, Eric says, the strangest aeroplane ever to grace the apron of A Shed at RAE Farnborough, where the experimental aircraft lived. Eric first flew it on 30 June 1944 and he says it was enjoyable but not really viable.

9. **Westland Whirlwind**. A good looker but with disappointing manoeuvrability and performance was how Eric would characterize this twin-engined fighter, which, perhaps with the right power plants, could have been a winner.

10. **Fairey Barracuda**. Although it was easy enough to deck land, Eric had mixed feelings about the big torpedo bomber. On 12 June 1943, when testing a radio altimeter fit, a test Barracuda lost power in the vicinity of Dunino airfield. In his logbook Eric records: 'flew through telephone wires and hit a tree; bounced; mowed down anti-invasion poles' before coming to rest. He describes it as a 'really shakey [sic] do'. But he and the flight test engineer survived, so it couldn't have been that bad!

Appendix 3: Winkle Firsts

First deck landing a jet. De Havilland Vampire (LZ551/G) at 11:28 on the flight deck of HMS *Ocean* on 3 December 1945, operating from RNAS Ford in Sussex. It is worth recording that Eric first flew Gloster E28/39 N4041/G on 29 May 1944 and reached Mach 0.82. His last flight was in EE215/G on 28 February 1945.

First deck landing of a twin-engined aircraft. On 25 March 1944, Eric became the first pilot to land a twin-engined aeroplane on an aircraft carrier in de Havilland Mosquito Mk VI LR359, which he described as a special Fleet Air Arm version but not yet a Sea Mosquito. The flight took fifty-five minutes to HMS *Implacable* from RNAS Crail. Eric completed eight deck landings in this Mosquito and a further seventeen flights in LR387 in May 1944. In his logbook, Eric records: 'the troops cheered; the officers smiled; the Civvies grinned, and the Flag Officer broke down sufficiently to say, "Thank you, nice work."'

First deck landing of twin-engined jet aircraft. Gloster Meteor on the flight deck of HMS *Implacable* on 8 June 1946.

In February 1948, Eric spent some time testing and evaluating the Sea Meteor for naval carrier duties but was firmly of the view that the engines took too long to spoil up in the event of the need to go ahead again if the arrester wires had been missed.

First deck landing of a tricycle. Bell Airacobra AH574 for deck landing assessment on 22 June 1944 at Farnborough before its one and only deck landing on 4 April 1945. The issue was initially the strength of the Airacobra to withstand arrester wire restraint and how best to catapult it back into the air.

First deck landing of a turbo-prop. Gloster Meteor conversion of a Mk III with MK IV parts and a Rolls-Royce turbo-prop engine. On 10 March 1948, Eric flew the Trent-engined fighter at Bitteswell, as the Admiralty believed that a turbo-prop might be better than an axial-flow jet engine for carrier-borne fighters.

First deck landing of a helicopter on a merchant ship. Eric first flew a Sikorsky Hoverfly from Speke on 4 March 1945, but his first official solo was 16 March 1945 in KK995. The same day he converted to the Gadfly version (KL107). The actual merchant-ship deck landing is not recorded in Eric's logbooks.

Appendix 4: Winkle's Aeroplanes

Eric claimed he had flown 487 types as pilot-in-command, but with good practice these types have been scrubbed by Dr Ron Hillary Smith of the Royal Aeronautical Society. He has arbitrated the detail which was taken from Eric's notes. His own record of types flown still stands at 487 even after scrutiny. Ron has deleted double counting of navalized versions and light aircraft definitions and could reduce that total to 480. However, I have allowed Sea Mosquito, Sea Vampire, Sea Meteor, Piper Grasshopper, AT-6 Harvard, P-40E and Tipsy Trainer to enter the list, as found in Eric's notes, to bring the list back to 487.

Aeroplanes

Aeronca Grasshopper

Aichi D3A Val

Airspeed Ambassador
Airspeed Envoy
Airspeed Horsa

Airspeed Oxford

Arado 96B
Arado 196A
Arado 199
Arado 232B
Arado 234B

Arado 240

Armstrong Whitworth Albemarle
Armstrong Whitworth Whitley

Auster Aiglet

Avro Anson
Avro Athena
Avro Lancaster
Avro Lancastrian
Avro Lincoln
Avro Manchester
Avro Shackleton
Avro Tudor
Avro Tutor
Avro York

BAC Jet Provost T3
BAC Lightning F1

BAE 125
BAE 146
BAE Hawk T1

Baynes Carrier Wing

Beagle B206 Basset
Beagle Pup

Beech 17 Traveller
Beech Baron
Beech Bonanza
Beech Super King Air

Beechcraft 18 Expediter

Bell Airacobra
Bell Airacomet
Bell King Cobra

Blackburn Beverley
Blackburn Botha
Blackburn Buccaneer
Blackburn Firebrand
Blackburn Firecrest
Blackburn Roc
Blackburn Shark
Blackburn Skua

Blohm & Voss 138
Blohm & Voss 141B
Blohm & Voss 222 Wiking
Blohm & Voss Bv 222

Boeing B-17G Fortress
Boeing B-29 Superfortress

Boulton Paul Defiant
Boulton Paul P108
Boulton Paul Sea Balliol

Bréguet Alizé
Bréguet Atlantique

Brewster Buffalo

Bristol Beaufighter
Bristol Beaufort
Bristol Blenheim Mk IV
Bristol Bombay
Bristol Brigand
Bristol Britannia
Bristol Buckingham
Bristol Bulldog
Bristol Freighter

British Aircraft Swallow

Britten-Norman Islander

Bücker Bestmann
Bücker Jungmann
Bücker Jungmeister
Bücker Student

Cant Z1007

Caproni 309
Caproni 311
Caproni Ca 135bis

Cessna 150
Cessna 177 Cardinal
Cessna 185 Skywagon
Cessna 195 Super Skymaster

Chance-Vought Corsair
Chance-Vought Cutlass

Chilton Aircraft DW1

Christlea Super Ace

Comper CLA7 Swift

Consolidated B-24 Liberator
Consolidated PBY-5 Catalina
Consolidated Vultee Privateer

Convair 240-5

Curtiss C-46 Commando
Curtiss P-36 Mohawk
Curtiss P-40E Kittyhawk
Curtiss P-40A Tomahawk
Curtiss SB2C Helldiver
Curtiss SO3C Seamew

Dassault Étendard
Dassault Mirage (III)
Dassault Mystère (IV)

De Havilland (Canada) Beaver
De Havilland (Canada) Chipmunk Mk 10
De Havilland Comet Mk 4
De Havilland Devon
De Havilland Heron
De Havilland DH 60 Gipsy Moth
De Havilland DH 80 Puss Moth
De Havilland DH 82 Tiger Moth
De Havilland DH 83 Fox Moth

De Havilland DH 85 Leopard Moth
De Havilland DH 86B Express
De Havilland DH 87B Hornet Moth
De Havilland DH 89 Rapide
De Havilland DH 93 Don
De Havilland DH 95 Flamingo
De Havilland Mosquito Mk IV/VI
De Havilland (Canada) Otter
De Havilland Sea Hornet
De Havilland Sea Mosquito
De Havilland Sea Vampire
De Havilland Sea Venom FAW 21
De Havilland Sea Vixen FAW 1
De Havilland Swallow
De Havilland Vampire

Dewoitine 520

DFS 230
DFS Kranich
DFS Weihe

Dornier Do 17
Dornier Do 18
Dornier Do 24
Dornier Do 26
Dornier Do 27
Dornier Do 217
Dornier Do 335

Douglas A-1 Skyraider
Douglas A-20 Boston
Douglas A-26 Invader
Douglas C-47 Dakota
Douglas F3D Skyknight
Douglas SDB Dauntless
Douglas TBD Devastator
Douglas C-54 Skymaster

Druine Turbulent

Elliott Newbury Eon

Embraer Bandeirante

English Electric Canberra

Erco Ercoupe Europa

Fairchild Argus
Fairchild Cornell
Fairchild XNQ-1

Fairey Albacore
Fairey Barracuda
Fairey Battle
Fairey Firefly
Fairey Gannet
Fairey Fulmar
Fairey Gordon
Fairey Seal
Fairey Spearfish
Fairey Swordfish
Fairey IIIF

Fiat BR20
Fiat C32
Fiat C42
Fiat G50

Fieseler Storch

Focke-Wulf 58
Focke-Wulf 189
Focke-Wulf 190
Focke-Wulf 200
Focke-Wulf Ta 152
Focke-Wulf Ta 154

Folland 43/37

Fouga Magister

Fournier (Sportavia) Milan

General Aircraft Cygnet
General Aircraft Hamilcar
General Aircraft Hotspur
General Aircraft L/56

Gloster E28/39
Gloster Gauntlet
Gloster Gladiator
Gloster Javelin
Gloster Meteor Mk I/III
Gloster Sea Meteor

Gotha 244

Grumman Ag-Cat
Grumman Albatross
Grumman Avenger
Grumman Bearcat
Grumman Cougar
Grumman Goose
Grumman Guardian
Grumman Hellcat
Grumman Panther
Grumman Tigercat
Grumman Widgeon
Grumman Wildcat Mk I/II/III/IV (Martlet)

Handley Page Gugnunc
Handley Page Halifax
Handley Page Halifax Mk II/IV
Handley Page Harrow
Handley Page Hastings
Handley Page Hermes
Handley Page Marathon

Hawker Fury
Hawker Hart
Hawker Hector
Hawker Henley
Hawker Hunter Mk F4/6
Hawker Hurricane Mk I/II
Hawker Nimrod
Hawker Osprey
Hawker P1040
Hawker P1052
Hawker P1127
Hawker Sea Fury FB10

Hawker Sea Hawk Mk 1/3/4/6
Hawker Tempest
Hawker Typhoon
Hawker Siddeley 748
Hawker Siddeley Gnat

Heinkel He 111
Heinkel He 115
Heinkel He 123
Heinkel He 129
Heinkel He 162
Heinkel He 177
Heinkel He 219

Henschel Hs 123
Henschel Hs 129

Heston Phoenix

Hitachi TR 2

Horton IV

Hunting Percival Jet Provost Mk I
Hunting Percival Provost

Ilyushin Il-2
Ilyushin Il-4

Jodel Ambassador
Jodel Club
Jodel Grand Tourisme
Jodel Mousqetaire
Jodel (SAN) Excellence
Jodel (SAN) Mascaret

Junkers Ju52-3M
Junkers Ju86
Junkers Ju87
Junkers Ju88
Junkers Ju188
Junkers Ju290
Junkers Ju352
Junkers Ju388

Kawasaki Ki-61 'Tony'

Klemm 35D
Klemm L25
Klemm L27

Lavochkin La-7

Le Vier Cosmic Wind

Lockheed C-130 Hercules
Lockheed Hudson
Lockheed F-80 Shooting Star
Lockheed F-104 Starfighter
Lockheed L-049 Constellation
Lockheed L-188 Electra
Lockheed P-2 Neptune
Lockheed P-38 Lightning

LTV F-8 Crusader

Luton Minor

Macchi C202
Macchi C205

Martin-Baker MB5
Martin Baltimore
Martin Marauder

McDonnell Banshee
McDonnell Douglas A-4 Skyhawk
McDonnell F-4N Phantom II

Messerschmitt Bf 108
Messerschmitt Bf 109
Messerschmitt Bf 110
Messerschmitt Me 163
Messerschmitt Me 262
Messerschmitt Me 410

Mikoyan-Gurevich MiG-3
Mikoyan-Gurevich MiG-15

Miles Aerovan
Miles Gemini
Miles Hawk Trainer
Miles Hobby
Miles Magister
Miles M.3 Falcon
Miles M.9 Master
Miles M.16 Mentor
Miles M.18 Trainer
Miles M.20 Lightweight Fighter
Miles M.25 Martinet
Miles M.28 Mercury
Miles M.33 Monitor
Miles M.35 Libellula
Miles M.38 Messenger
Miles M.38 Messenger 3
Miles Mohawk
Miles Monarch
Miles Sparrowhawk

Mitsubishi G4M 'Betty'
Mitsubishi Ki-46 'Dinah'
Mitsubishi A6M 'Zeke'

Mooney M20

Morane-Saulnier 406
Morane-Saulnier Paris
Morane-Saulnier Rallye

Muntz Youngman-Baynes

Nakajima Ki-43 'Oscar'
Nakajima Ki-84 'Frank'

Nationalsozialistisches Fliegerkorps SG38

(Tipsy) Nipper III

Noorduyn Norseman

Nord 262A
Nord Noralpha
Nord Pingouin

North American A-2 Savage
North American AT-6 Harvard
North American B-25 Mitchell
North American F-86 Sabre
North American F-100 Super Sabre
North American P-51 Mustang
North American T-6 Texan

Northrop 24 Gama
Northrop F-5 Freedom Fighter
Northrop P-61 Black Widow

Orlikan Meta Sokol

Percival Gull
Percival Pembroke
Percival Prentice
Percival Proctor
Percival Q6
Percival Vega Gull

Petlyakov PE-2

Piaggio P136
Piaggio P166

Piel Emeraude

Pilatus Porter

Piper Cub
Piper Cub Special 90
Piper L-4 Grasshopper
Piper PA-12 Supercruiser
Piper PA-22 Tripacer
Piper PA-23 Apache
Piper PA-24 Comanche
Piper PA-25 Pawnee
Piper PA-27 Aztec
Piper PA-28 Cherokee
Piper PA-31 Navajo
Piper PA-34 Seneca

Pitts S2 Special

Polikarpov I-15
Polikarpov I-16

Portsmouth Aerocar Major

Reggiane 2000 Falco
Reggiane 2001

Reid & Sigrist Desford

Republic F-84 Thunderjet
Republic F-84F Thunderstreak
Republic P-43 Lancer
Republic P-47 Thunderbolt
Republic Seabee

Robin HR100 Royale

Rollason D62 Condor

Ryan FR Fireball

Saab 32 Lansen
Saab 91 Safir
Saab 105
Saab J21
Saab J29 Tunnan

Saunders-Roe SR/A1

Savoia-Marchetti SM70
Savoia-Marchetti SM82
Savoia-Marchetti SM95

Scheibe Motorspatz

Schmetz Olympia-Meise

Schneider Grunau Baby

Scottish Aviation Bulldog
Scottish Aviation Pioneer
Scottish Aviation Twin Pioneer

Short S31
Short Sealand
Short Skyvan
Short Stirling
Short Sturgeon

SIAI-Marchetti SF 260

Siebel Si-204

Sipa S903

Slingsby T.7 Kirby Cadet
Slingsby T.21 Sedbergh
Slingsby T.30 Prefect
Slingsby T.31 Tandem Tutor
Slingsby T.49 Capstan
Slingsby T.45 Swallow
Slingsby T.61 Motor Tutor

SOCATA ST-10 Diplomate

Stampe & Vertongen SV-4

Stearman 75 Kaydet

Stinson Junior R
Stinson Reliant
Stinson Sentinel

Supermarine S24/37 Type 322
Supermarine Attacker
Supermarine Seafang
Supermarine Seafire
Supermarine Seagull
Supermarine Sea Otter
Supermarine Scimitar
Supermarine Spiteful
Supermarine Spitfire
Supermarine Walrus

Szybowcowy Zakład Doświadczalny SZD-9 Bocian

Taylor JT1 Monoplane
Taylor JT2 Titch

Taylorcraft Auster

Thruxton Jackaroo

Tipsy S2
Tipsy Trainer
Tipsy Type B

Vickers Valiant
Vickers Vanguard
Vickers VC10
Vickers Viking
Vickers Viscount
Vickers Warwick
Vickers Wellington
Vickers Windsor

Vought-Sikorsky Chesapeake

Vought-Sikorsky Kingfisher

Vultee Vengeance

Waco CG-3
Waco Hadrian

Westland Lysander
Westland Welkin
Westland Whirlwind
Westland Wyvern

Winter Braunschweig LF-1 Zaunkönig (Wren)

Yakovlev Yak-1
Yakovlev Yak-9
Yakovlev Yak-11

Zlin Akrobat

Helicopters

Aérospatiale Alouette III
Aérospatiale Ecureuil (Squirrel)
Aérospatiale Twin Squirrel

Agusta A109

Bell 47G
Bell 204
Bell 222

Bell AH-1
Bell HTL-5
Bell JetRanger
Bell LongRanger

Boeing Vertol Chinook

Brantly B-2

Bristol Sycamore

Enstrom F28
Enstrom Shark

Hillier UH-12B THE

Hughes 300
Hughes 500

Kamov 26

MBB Bo 105

Mil Mi-1
Mil Mi-2
Mil Mi-4

Piasecki Retriever

Robinson R-22

Saunders-Roe P531

Saunders-Roe Skeeter

Sikorsky R-4B Hoverfly
Sikorsky R-6 Hoverfly II
Sikorsky S-55 HRS
Sikorsky S-58T
Sikorsky S-61
Sikorsky S-76

Sud-Aviation Djinn

Vertol 107

Westland-Aérospatiale Gazelle
Westland-Aérospatiale Lynx
Westland-Sikorsky S-51 Dragonfly
Westland-Sikorsky S-55 Whirlwind
Westland Wasp
Westland Wessex Mk 1/3/5

Acknowledgements

This has been a labour of love and a voyage of exploration in search of the real Eric Brown. It was clear from the beginning that it would be a longer search than I had planned and that I would need to seek the help and guidance of friends, old and new. So, I raided my address books from forty years of writing about the Fleet Air Arm and the records which Eric himself had left. Glenn Melrose-Brown was as keen as I to find out the back story and to understand some of the myths and stories which he had heard and didn't quite believe. I also found James Melrose and Margaret Scott, descendants of Euphemia Melrose Brown in Galashiels, and spent some delightful hours with them. Without the family's support and those at the Fleet Air Arm, some sadly departed since the book was commissioned, this biography would not have been possible, and the back story would not have been retrieved. To everyone who has contributed, warts and all, my undying thanks and gratitude. It is important to remember and thank the late Jean Kelly, Eric's companion for his last seventeen years, who gave him a new focus and the strength and encouragement to keep going.

Thanks go firstly to Matthew Wills, the naval aviation historian who has been my conscience and proofreader. Having someone who knows the naval aviation scene and who is a professional editor allowed me to deliver the first drafts with greater confidence than I would have expected.

Unravelling Eric's back story took time – twelve months longer than planned – and would not have been possible without the record departments of the Royal Navy at HMS *Excellent* and the Royal Air Force at RAF Cranwell. Having Eric's S 206 official record, immediately put me on the right track and gave me the confidence to contradict some of the previous perceived history, much of it stemming from *Wings on My Sleeve*.

Getting the local history and knowledge of people in the Borders was very important. This is where Ian Brown (National Museum of Scotland) and Cathy Tucker (Royal Scots Museum) helped. The birth certificate issue, dealing with Eric's year and place of birth in particular, could not have been tackled without Heather McMillan and the Clackmannanshire Registration Office team. Reg Pettit and other family friends and relations offered titbits for the better understanding of Eric's complex relationship with his Scottish homes. The official recorders at the Royal High School and the University of Edinburgh archives confirmed dates and times for me. Again, thank you.

In Germany, Brigitte Mohn, archivist and cultural scientist at the Kurt-Hahn-Archiv in the Kreisarchiv Bodenseekreis und Kreiskulturamt laid low some of the previously presumed facts about 1939. Professor Peter Caddick-Adams and Rob Schaefer helped me understand the Nazi state security system, and several members of the Special Forces Club were illuminating on the way in which the Secret Intelligence Service/MI6 worked in the pre-war, wartime and then Cold War periods, when Eric was interested in Germany. Dr Jann de Witt at the Deutchesmarinebund provided excellent reference material. George Gebauer, a close friend of my cousin Margaret Bennett, was actually a young boy in Berlin in the 1930s, so his memories have helped me record what Eric would have seen at the Olympic Games. Clare Mulley, author and lecturer, helped me understand the relationship which Eric had with Hanna Reitsch and stopped me getting too carried away. The Hon. Dr Katharine Campbell, the daughter of Lord (Sholto) Douglas, helped with insights into Germany 1945 and the state of mind of those who have witnessed or experienced great trauma.

Poignant too was reading the letters of Norris Patterson, a young naval pilot who went down with HMS *Audacity* in 1941 but whose sister, the late Mary Sturgess (née Patterson), kept his memory very much alive. They helped me shape the shipboard life and Eric's attitude to life at the time. I was amazed to find that some of Eric's messmates and eyewitnesses to some of his greatest feats of aviation were still alive and willing to talk to me. Ralph Jameson, Ian Richardson, John Fay and Captain Colin Robinson were there and saw the first twin-engined aeroplane, the first jet land and the use of a rubber deck on aircraft carriers. Nick Cook, Eric's Sea Hawk wingman, Rear-Admiral Mike Layard, Mike Tristram, the late Rear-Admiral Colin Cook-Priest, the late Lieutenant-Commander David Howard, Kevin Sharman and the late Dave (Shorty) Hamilton were robust in their feelings, but their views are nevertheless valid. Thanks to Robin Spratt, Jilly Wise and Gill Kerslake of the Fleet Air Arm Officers' Association for the introductions. Colour was provided by Ivor Faulkner, a flying boat pilot in the war with perfect recall, and David Gibbings helped with the Vampire notes.

When I needed some help with the Martlet/Wildcat, I turned to Dave Southwood, former senior tutor at the Empire Test Pilots' School, who has flown the later marks, known as the Wildcat, and had discussed flying with Eric. The young Eric of the period has been faithfully reproduced by the sculpture of Jenna Gearing, whose recollections of visits to, and conversations with, Eric were most helpful and always involved a glass of something.

Eric's Farnborough days were some of the happiest in his life, if not the happiest. To find and speak at length with Geoffrey Cooper, at over 100 years old the last 'boffin', was a real pleasure. With the help of Nicholas Jones at Quanta Films, we have captured some of his memories on film. Geoffrey was even once human ballast for a test flight piloted by Eric. We were hosted there by Richard Gardner and the Farnborough Air Sciences Trust, one of Eric's favourite organizations and one which treasures his association with Farnborough.

Edwin (Eddie) Albrecht explained the fascination both he and Eric had with big band sounds and the Glenn Miller connection. Geoff Dunford, who turns out to have been a childhood friend of my favourite uncle, was an eyewitness to some of Eric's glider-flying; his contact came to me from the Clatford Village Shop, my local post office. Small world.

Postwar flying with Sea Fury and Sea Hawk in Germany and Eric's post-career with the British Helicopter Advisory Board would not have been completed without Brian (Schmoo) Ellis, the reluctant hero of Fleet Air Arm air operations over Korea. Stories of squadron life have been told and faithfully recorded.

Brian Riddle at the National Aerospace Library provided useful references and leads, as did the staff of the Fleet Air Arm Museum, who allowed me to pore over the official record books and line books of 802 Naval Air Squadron; sadly, those of 801 Squadron went down with HMS *Audacity* along with many other records which would have been so useful. When the family parted with Eric's medals, Bonhams provided the venue and John Millensted had the expert knowledge of medals.

A long weekend travelling to Bonn with Glenn and his childhood friends opened up a new vista of Eric's life as naval attaché in the Cold War. Huge thanks to Ros Melrose-Brown, Isobel Trowers, David and Jennie Franks, Veronica

Bryant, Michael and Jane May for sharing their trip with the Beaver family.

At Copthorne, during the funeral and wake, and then at Yeovilton for the memorial service Emily Horton held the event together, and people with great stories came to talk to me. Thank you, even if your name isn't mentioned, your help is remembered. Old friends of Eric and of mine: Admiral Sir George Zambellas, Vice-Admiral Keith Blount, Rear-Admiral Russ Harding, Commodore Jock Alexander and his team at Navy Wings, including Sue Eagles, Laila Salter, Jon Parkinson, together with Fly Navy Heritage Trust ambassadors Alfie Southwell, Tim Manna and Stephen Partridge-Hicks, who are so generous to the memory of Winkle. A writer should not delve into the aviation history of Farnborough without speaking with Sir Gerald Howard, the former minister and its Member of Parliament for several decades; he also helped with some important Whitehall-related points which needed clarification. The late Group Captain Don McClen was kind with his thoughts, advice and memories, especially of his mentor and Eric's sometime nemesis, Admiral Sir Raymond Lygo.

Airbus, Europe's aerospace and defence giant, is one of the key organizations that recognized Eric's merit, and it named its VIP Lounge at Toulouse in his memory. Thanks to Jeremy Greaves, Hans-Ulrich Willbold and the team for hosting the Melrose-Browns and me in France in 2016 and me again at the Farnborough Air Show that year.

Eric would have been thrilled that France and Germany pay as much attention to him as the Americans. Germany always loomed large in Eric's life so, after his death, it was good to make contact with Peter Selinger, who met Eric when Professor Dr Karl Nickel, husband of Gunilde Horten, celebrated the publication of the book *Tailless Aircraft in Theory and Practice* in the English language, for which Eric kindly worked as translator and lecturer on 4 September 1994 at Freiburg/Breisgau. Before Gunilde passed away in the summer of 2016, the last letter she wrote was to me about Eric.

Three more people must be mentioned: Commander Barney Wainwright, Wing Commander John Davis and Matthew Wills. They all corrected, amended, proofed, commented and advised me on the text, syntax and context. I have been advised, encouraged and sometimes even flattered by James Holland, Trevor Dolby and Michael Ivey.

Finally, the hard work, friendship and advice of Rowland White, himself a great author as well as editor and publisher, made my job easier. A superb edit and 'slim-down' was faultlessly performed by Eleo Gordon, and I know how lucky I was to have her guidance. My agent, Victoria Hobbs at AM Heath, is a wealth of guidance and a great negotiator. I am grateful to the team at Michael Joseph: Jorgie Bain (editorial assistant), Ruth Atkins (assistant editor), Nick Lowndes (managing editor), David Watson (copy editor) and Lauren Wakefield (designer). Thank you all.

Finally, without the constant support of my wife Cate and son Jack, both pilots and keen historians, this would not have happened. My love and gratitude to them. As usual, the errors and omissions are my fault alone.

Abbreviations

A&AEE	Aeroplane and Armaments Experimental Establishment
Bf	Bayerischer Flugzeugwerke (German aircraft manufacturer, later Messerschmitt)
BHAB	British Helicopter Advisory Board
CBE	Companion of the British Empire
EA	Enemy aircraft
Fw	Focke-Wulf (German aircraft manufacturer)
HCGB	Helicopter Club of Great Britain
HMS	His/Her Majesty's Ship
MBE	Member of the British Empire
Me	Messerschmitt (German aircraft manufacturer)
OBE	Order of the British Empire
RAE	Royal Aircraft Establishment (Farnborough) (Bedford)
RAF	Royal Air Force
RAFVR	Royal Air Force Volunteer Reserve
RFC	Royal Flying Corps
RN	Royal Navy
RNAS	Royal Naval Air Station

RNR	Royal Naval Reserve
RNVR	Royal Naval Volunteer Reserve
STU	Service Trials Unit
USAAC	United States Army Air Corps
USAAF	United States Army Air Force
USAF	United States Air Force
USN	United States Navy

Bibliography

Beaver, Paul, *The British Aircraft Carrier* (PSL, 1982, 1984, 1986)

Beaver, Paul, *Encyclopaedia of the Modern Fleet Air Arm* (PSL, 1984)

Beaver, Paul, *Winkle Tribute* (BWL, 2016, 2017)

Beaver, Paul, *Forgotten Few* (BWL, 2019)

Beevor, Antony, *The Battle for Spain* (Phoenix, 2007)

Brown, Eric Melrose and Bancroft, Dennis, *Miles M52* (Spellmount, 2012)

Brown, Eric Melrose, *Too Close for Comfort: One Man's Close Encounters of the Terminal Kind* (Blacker, 2015)

Brown, Eric Melrose, *Wings on My Sleeve* (various)

Brown, Eric Melrose, *Wings of the Luftwaffe* (Hikori, 1980, 2013)

Brown, Eric Melrose, *Wings of the Navy* (Hikori, 1980, 2013)

Brown, Eric Melrose, *Wings of the Weird and Wonderful* (Hikori 1980, 2013)

Churchill, Winston, *The Second World War: Their Finest Hour* (Folio edition, 2000)

Cooper, Geoffrey, *Farnborough and the Fleet Air Arm* (Midland Counties, 2008)
Cull, Brian, *Battle for the Channel* (Fonthill, 2017)
Fay, John, *Golden Wings and Navy Blue* (Hallmark, 2008)
Fleet Air Arm Officers' Association, *The Song Book* (FAAOA, 2010)
Gebauer, George, *Hitler Youth to Church of England Priest* (Createspace, 2014)
Henshaw, Alex, *Sigh for a Merlin* (AirData, 1992)
Herlin, Hans, *Udet – A Man's Life* (MacDonald, 1960)
Hilmes, Oliver, *Berlin 1936* (Bodley Head, 2016)
HMSO, *The Fleet Air Arm, the Admiralty Account of Naval Air Operations* (HMSO, 1943)
Jackson, A. J., *De Havilland Aircraft* (Putnam, 1987)
Killen, John, *The Luftwaffe, a History* (Pen & Sword, 1967)
Longden, Sean, *T-Force* (Constable, 2009)
Lygo, Raymond, *Collision Course* (Book Guild, 2002)
Mulley, Clare, *The Women Who Flew for Hitler* (Macmillan, 2017)
Penrose, Harold, *No Echo in the Sky* (Fonthill, 2016)
Quill, Jeffrey, *Spitfire* (various)
Reitsch, Hanna, *The Sky, My Kingdom* (Greenhill, 1991)
Sturdivant, Ray, *Squadrons of the Fleet Air Arm* (Air Britain, 1984)
Thetford, Owen, *British Naval Aircraft Since 1921* (Putnam, 1958)

Turner, Barry, *Karl Dönitz and the Last Days of the Third Reich* (Icon, 2016)

Wellum, Geoffrey, *First Light* (Penguin, 2003)

White, Rowland, *Into the Black* (Bantam Press, 2016)

Wood, Derek, *Project: Cancelled* (Jane's, 1987)

Wragg, David, *Swordfish* (Weidenfeld & Nicolson, 2003)

Eric's notes, photographs and sketches have been used throughout this book. I was also lucky enough to hear Eric speak over 200 times at public and private events. In his last years, mentally preparing for this book, I spent time with Eric in person or on the telephone. My notes and memories are reflected in this book.

Picture Permissions

Every effort has been made to ensure images are correctly attributed, however if any omission or error has been made, please notify the publisher for correction in future editions.

Integrated pictures

BBC: p.424
Brett Carlton iStock via Getty Images: p.168
Interfoto / Alamy Stock Photo: pp.12, 242
Michael Turner: p.70
NASA: p.354
Philip E. West, www.philipewest.co.uk: p.152
Public Domain: pp.446, 449, 451, 454, 455, 460, 461, 462, 463, 465, 467
Stocktrek images via Getty: p.302
Unknown: pp.30, 192, 224, 270
USAF: p.447
Wikimedia Commons: pp.110, 370, 444, 445, 448, 450, 452, 453, 456, 458, 459, 464, 466
Winkle Archive: pp.xx, 2, 40, 54, 62, 80, 98, 136, 178, 200, 292, 324, 338, 386, 404

Inset pictures

Air Collection 216 / Alamy Stock: p.14, top
BBC: p.24, middle
Crown Copyright: p.2, bottom right; p.3, bottom right; p.4, middle right; p.12, top right; p.19, middle right; p.21, bottom right
Georg Pahl: p.2, top left
Interfoto / Alamy Stock Photo: p.12, bottom
Jon Freeman (artist) and Beaver Westminster Ltd: p.10, middle left; p.13, 3rd row
Ladybird Books Ltd: p.7; p.14, middle
Michael Fader, http://www.wings-aviation.ch/: p.4, middle left
Michael Turner: p.8, top; p.9, bottom; p.11 top; p.14, bottom
NASA: p.17, top right, middle
National Portrait Gallery, London: p.13, 2nd row left
Salvation Army International Heritage Centre: p.1, top
Unknown: p.16, top right; p.24 top left
Wikimedia Commons: p.2, middle right; p.5 top; p.12, middle; p.16, top left; p.17, top left; p.21, top right
Winkle Archive: p.1, middle left & right, bottom; p.2, top right, middle, bottom left; p.3, top, middle, bottom left; p.4, top, bottom; p.5, middle left & right, bottom; p.6, all; p.8, bottom; p.9, top left & right; p.10, top, bottom; p.11, middle, bottom; p.13, top, 2nd row right, 4th row, 5th row; p.15, all; p.16, middle, bottom left & right; p.17 bottom; p.18, all; p.19, top, bottom; p.20, all; p.21, top left, middle, bottom left; p.22, all; p.23, all; p.24, bottom

Index